SPACE SCIENCE, EXPLORATION AND POLICIES

NASA

MAJOR PROJECTS AND SELECT ISSUES

SPACE SCIENCE, EXPLORATION AND POLICIES

Additional books and e-books in this series can be found
on Nova's website under the Series tab.

SPACE SCIENCE, EXPLORATION AND POLICIES

NASA

MAJOR PROJECTS AND SELECT ISSUES

OLIVER CUMMINGS
EDITOR

snova
New York

Library of Congress Cataloging-in-Publication Data

ISBN: 978-1-53614-817-6

Published by Nova Science Publishers, Inc. † New York

CONTENTS

PREFACE

NASA's mission is to drive advances in science, technology, aeronautics, and space exploration to enhance knowledge, education, innovation, economic vitality, and stewardship of Earth. Chapter 1 and 2 provide a snapshot for 2018 of how well NASA is planning and executing its major acquisition projects. In May 2017, GAO found that projects were continuing a generally positive trend of limiting cost and schedule growth, maturing technologies, and stabilizing designs. But, at the same time, GAO noted that many of these projects, including some of the most expensive ones, were approaching the phase in their life cycles when cost and schedule growth is most likely. Chapter 3 is an overview of the 2017 budget. Chapter 4 determines the extent to which NASA has established and implemented leading IT management practices in strategic planning, workforce planning, governance, and cybersecurity. Chapter 5 addresses, among other things, the extent to which (1) NASA's Inspector General investigated contractor and grantee whistleblower reprisal complaints; (2) NASA's Administrator reviewed reprisal complaints in a timely manner; and (3) NASA communicated the applicable whistleblower reprisal protections to contractors. Chapter 6 sets forth policy governing the release of public information, which is defined as information in any form provided to news and information media, especially information that has the potential to generate significant media, or public interest or inquiry.

Chapter 1 - This report provides GAO's annual snapshot for 2018 of how well NASA is planning and executing its major acquisition projects. In May 2017, GAO found that projects were continuing a generally positive trend of limiting cost and schedule growth, maturing technologies, and stabilizing designs. But, at the same time, GAO noted that many of these projects, including some of the most expensive ones, were approaching the phase in their life cycles when cost and schedule growth is most likely.

The explanatory statement of the House Committee on Appropriations accompanying the Omnibus Appropriations Act, 2009 included a provision for GAO to prepare status reports on selected large-scale NASA programs, projects, and activities. This is GAO's

10th annual assessment. This report describes the cost and schedule performance of NASA's portfolio of major projects, among other issues. This report also includes assessments of NASA's 26 major projects, each with a life-cycle cost of over $250 million. To conduct its review, GAO analyzed cost, schedule, technology maturity, design stability, and other data; reviewed monthly project status reports; and interviewed NASA officials.

Chapter 2 - GAO designated NASA's acquisition management as a high-risk area in 1990 after a history of persistent cost growth and schedule slippage in many of NASA's major projects. In more recent years, GAO found that NASA had taken some steps to improve its management, and, in May 2017, GAO found that projects were continuing a generally positive trend of limiting cost and schedule growth. But at the same time, GAO noted that many of these projects, including some of the most expensive ones, were approaching the phase in their life cycles when cost and schedule growth is most likely.

This statement summarizes GAO's 2018 findings from its 10th annual snapshot of how well NASA is planning and executing its major acquisition projects, and describes (1) the cost and schedule performance of NASA's portfolio of major projects and (2) the extent to which NASA faces risks for further cost increases and schedule delays. To conduct its review for the 2018 report, GAO-18-280SP, GAO analyzed cost, schedule, and other data for NASA's 26 major projects, each with a life-cycle cost of over $250 million; reviewed monthly project status reports; and interviewed NASA officials.

Chapter 3 - The National Aeronautics and Space Administration (NASA) was created in 1958 by the National Aeronautics and Space Act (P.L. 85-568) to conduct civilian space and aeronautics activities. It has four mission directorates. The Science Mission Directorate manages robotic science missions, such as the Hubble Space Telescope, the Mars rover Curiosity, and satellites for Earth science research. The Aeronautics Research Mission Directorate conducts research and development on aircraft and aviation systems. The Space Technology Mission Directorate develops technologies for use in future space missions, such as advanced propulsion and laser communications. The Human Exploration and Operations Mission Directorate is responsible for human spaceflight activities, including the International Space Station and development efforts for future crewed spacecraft. In addition, NASA's Office of Education manages formal and informal education programs for school children, college and university students, and the general public.

While Congress is generally supportive of most NASA programs, government-wide fiscal constraints make funding decisions challenging. The Administration has requested $19.025 billion for NASA in FY2017. This amount is 1.3% less than the FY2016 appropriation of $19.285 billion. Unusually, the FY2017 request includes $763 million in mandatory funds. The House bill (H.R. 5393) would provide $19.508 billion. The Senate bill (S. 2837) would provide $19.306 billion. Neither bill includes mandatory funding.

The FY2017 request for the Science Mission Directorate is $5.601 billion, an increase of 0.2% from FY2016. Within this total, funding for Earth Science, Astrophysics, and Heliophysics would increase, while funding for Planetary Science and the James Webb

Space Telescope would decrease. The House bill would provide $5.597 billion for Science, while the Senate bill would provide $5.395 billion. Within these totals, the bills differ considerably in their allocation of funding between Earth Science and Planetary Science.

The FY2017 request for the Aeronautics Research Mission Directorate is $790 million, an increase of 23.5% from FY2016. The request includes New Aviation Horizons (NAH), a new initiative of experimental aircraft and systems demonstrations. The House and Senate bills would provide $712 million and $601 million, respectively, for Aeronautics.

The FY2017 request for the Space Technology Mission Directorate is $827 million, an increase of 20.4% from FY2016. The House and Senate bills would provide $739 million and $687 million.

For the Human Exploration and Operations Mission Directorate, the FY2017 request for Exploration is $3.337 billion, a decrease of 17.2% from FY2016, while the request for Space Operations is $5.076 billion, an increase of 0.9%. The Exploration request includes $1.263 billion, a decrease of 35.2%, for Space Launch System launch vehicle development. Funding for the Commercial Crew program (formerly requested in Exploration) is combined with funding for operational cargo and crew transport to the International Space Station in a new Space Transportation item within Space Operations. The House bill would provide $4.183 billion for Exploration, including $2.000 billion for SLS development, and $4.890 billion for Space Operations. The Senate bill would provide $4.330 billion for Exploration, including $2.150 billion for the SLS, and $4.951 billion for Space Operations.

The FY2017 request for the Office of Education is $100 million, a decrease of 13.0% from FY2016. The House and Senate bills would provide $115 million and $108 million, respectively. The request would reduce funding for the National Space Grant College and Fellowship Program, the Experimental Program to Stimulate Competitive Research, and the Minority University Research Education Program. Both bills would fund these programs at their FY2016 levels.

Chapter 4 - NASA depends heavily upon IT to conduct its work. The agency spends at least $1.5 billion annually on IT investments that support its missions, including ground control systems for the International Space Station and space exploration programs.

The National Aeronautics and Space Administration Transition Authorization Act of 2017 included a provision for GAO to review the effectiveness of NASA's approach to overseeing and managing IT, including its ability to ensure that resources are aligned with agency missions and are cost effective and secure. Accordingly, GAO's specific objective for this review was to determine the extent to which NASA has established and implemented leading IT management practices in strategic planning, workforce planning, governance, and cybersecurity. To address this objective, GAO compared NASA IT policies, strategic plans, workforce gap assessments, and governance board documentation to federal law and leading practices. GAO also assessed NASA IT security plans, policies, and procedures against leading cybersecurity risk management practices.

Chapter 5 - NASA obligated over $18 billion in contracts and more than $1 billion in grants in fiscal year 2016, and it relies on a significant number of contractor and grantee employees to accomplish its work. These employees are legally protected from reprisal—such as demotion or firing—for whistleblowing.

GAO was asked to review NASA's whistleblower reprisal protections for contractor and grantee employees. This report addresses, among other things, the extent to which (1) NASA's Inspector General investigated contractor and grantee whistleblower reprisal complaints; (2) NASA's Administrator reviewed reprisal complaints in a timely manner; and (3) NASA communicated the applicable whistleblower reprisal protections to contractors. GAO reviewed NASA and its Inspector General's policies and guidance; reviewed a generalizable sample of 100 contracts from all NASA centers with contracts in fiscal year 2016; and interviewed relevant officials and contractors, grantees, and advocacy groups.

Chapter 6 - This directive sets forth policy governing the release of public information, which is defined as information in any form provided to news and information media, especially information that has the potential to generate significant media, or public interest or inquiry. Examples include, but are not limited to, press releases, media advisories, news features, and web postings. Not included under this definition are scientific and technical reports, web postings designed for technical or scientific interchange, and technical information presented at professional meetings or in professional journals.

In: NASA: Major Projects and Select Issues
Editor: Oliver Cummings
ISBN: 978-1-53614-817-6

Chapter 1

NASA: ASSESSMENTS OF MAJOR PROJECTS*

United States Government Accountability Office

WHY GAO DID THIS STUDY

This report provides GAO's annual snapshot for 2018 of how well NASA is planning and executing its major acquisition projects. In May 2017, GAO found that projects were continuing a generally positive trend of limiting cost and schedule growth, maturing technologies, and stabilizing designs. But, at the same time, GAO noted that many of these projects, including some of the most expensive ones, were approaching the phase in their life cycles when cost and schedule growth is most likely.

The explanatory statement of the House Committee on Appropriations accompanying the Omnibus Appropriations Act, 2009 included a provision for GAO to prepare status reports on selected large-scale NASA programs, projects, and activities. This is GAO's 10th annual assessment. This report describes the cost and schedule performance of NASA's portfolio of major projects, among other issues. This report also includes assessments of NASA's 26 major projects, each with a life-cycle cost of over $250 million. To conduct its review, GAO analyzed cost, schedule, technology maturity, design stability, and other data; reviewed monthly project status reports; and interviewed NASA officials.

* This is an edited, reformatted and augmented version of United States Government Accountability Office; Report to Congressional Committees, Publication No. GAO-18-280SP, dated May 2018.

What GAO Recommends

In prior reports, GAO has made related recommendations that NASA generally agreed with, but has not yet fully addressed. GAO continues to believe they should be fully addressed. NASA generally agreed with GAO's findings.

What GAO Found

The cost and schedule performance of the National Aeronautics and Space Administration's (NASA) portfolio of major projects has deteriorated, but the extent of cost performance deterioration is unknown. NASA expects cost growth for the Orion crew capsule—one of the largest projects in the portfolio—but does not have a current cost estimate. In addition, the average launch delay for the portfolio was 12 months, the highest delay GAO has reported in its 10 years of assessing major NASA projects (see figure below).

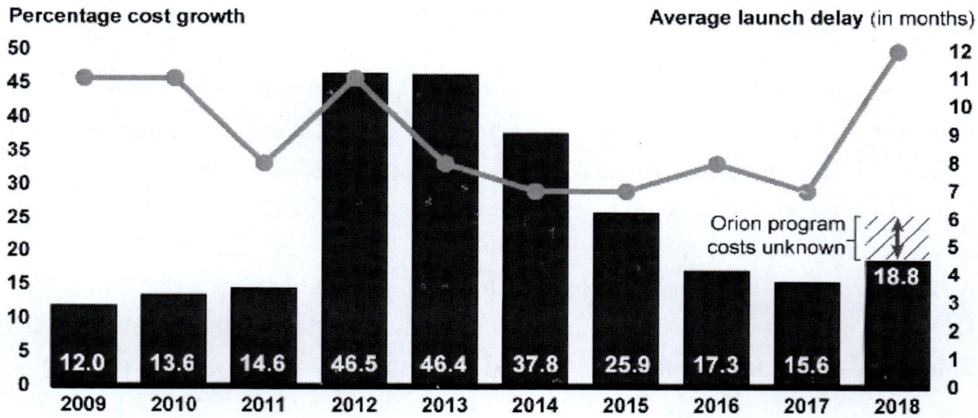

Source: GAO analysis of National Aeronautics and Space Administration data. | GAO-18-280SP.

NASA's Major Project Portfolio Cost and Schedule Performance Deteriorated in 2018.

The deterioration in portfolio performance was the result of 9 of the 17 projects in development experiencing cost or schedule growth.

- Four projects encountered technical issues that were compounded by risky program management decisions. For example, the Space Launch System and Exploration Ground Systems programs are large-scale, technically complex human spaceflight programs, and NASA managed them to aggressive schedules and with insufficient levels of cost and schedule reserves. This made it more difficult for the programs to operate within their committed baseline cost and schedule estimates.

- Two projects ran into technical challenges that resulted in delays in the integration and test phase. For example, in December 2017, GAO found that the James Webb Space Telescope project encountered delays primarily due to the integration of the various spacecraft elements taking longer than expected, as well as resolving technical issues during testing. GAO has previously found that integration and testing is when projects are most at risk of incurring cost and schedule growth.
- Three projects experienced cost growth or schedule delays due to factors outside of the projects' control, such as delays related to their launch vehicles.

NASA continues to face increased risk of cost and schedule growth in future years due to new, large and complex projects that will enter the portfolio and expensive projects remaining in the portfolio longer than expected.

ABBREVIATIONS

AIM	Asteroid Impact Monitor
ATLAS	Advanced Topographic Laser Altimeter System
CCP	Commercial Crew Program
CNES	Centre National d'Etudes Spatiales
DART	Double Asteroid Redirection Test
DCI	data collection instrument
DSOC	Deep Space Optical Communications
EGS	Exploration Ground Systems
EM-1	Exploration Mission 1
EM-2	Exploration Mission 2
ESM	European Service Module
ESA	European Space Agency
GFAS	Ground Flight Application Software
GFZ	German Research Centre for Geosciences
GN&C	Guidance, Navigation, and Control
GRACE-FO	Gravity Recovery and Climate Experiment Follow-On
GSLV	Geosynchronous Satellite Launch Vehicle
HP3	Heat Flow and Physical Properties Probe
HVPS	high voltage power supply
ICESat-2	Ice, Cloud, and Land Elevation Satellite-2
ICON	Ionospheric Connection Explorer

InSight	Interior Exploration using Seismic Investigations, Geodesy, and Heat Transport
ISRO	Indian Space Research Organisation
ISS	International Space Station
JCL	joint cost and schedule confidence level
JPL	Jet Propulsion Laboratory
JPSS-2	Joint Polar Satellite System 2
JWST	James Webb Space Telescope
KaRIn	Ka-band Radar Interferometer
KDP	key decision point
LBFD	Low Boom Flight Demonstrator
LCRD	Laser Communications Relay Demonstration
LRI	Laser Ranging Interferometer
MAVEN	Mars Atmosphere and Volatile Evolution
NASA	National Aeronautics and Space Administration
NEXT-C	NASA's Evolutionary Xenon Thruster-Commercial
NISAR	NASA ISRO Synthetic Aperture Radar
NOAA	National Oceanic and Atmospheric Administration
NPR	NASA Procedural Requirements
OCI	Ocean Color Instrument
Orion	Orion Multi-Purpose Crew Vehicle
OSIRIS-REx	Origins-Spectral Interpretation-Resource Identification Security-Regolith Explorer
PSP	Parker Solar Probe
PACE	Plankton, Aerosol, Cloud, ocean Ecosystem
QueSST	Quiet Supersonic Technology
RBI	Radiation Budget Instrument
SCaN	Space Communication and Navigation
SCCS	Spaceport Command and Control System
SCS	Sampling and Caching System
SDO	Solar Dynamics Observatory
SEIS	Seismic Experiment for Interior Structure
SGSS	Space Network Ground Segment Sustainment
SLS	Space Launch System
SPC	Solar Probe Cup
SWOT	Surface Water and Ocean Topography
TESS	Transiting Exoplanet Survey Satellite
VBB	Very Broad Band
WFIRST	Wide-Field Infrared Survey Telescope
WICN	Workforce Information Cubes for NASA

BACKGROUND

The National Aeronautics and Space Administration (NASA) is planning to invest about $61 billion over the life cycle of its current portfolio of 26 major projects—those projects that have a life-cycle cost over $250 million. These projects aim to continue exploring Earth and the solar system and extend human presence beyond low Earth orbit, among other things. This report provides an overview of NASA's planning and execution of these major acquisitions—an area that has been on GAO's high risk list since 1990.[1] It includes assessments of NASA's key projects across mission areas, such as the Space Launch System for human exploration, Mars 2020 for planetary science, and Ice, Cloud, and Land Elevation Satellite-2 (ICESat-2) for Earth science.

The explanatory statement of the House Committee on Appropriations accompanying the Omnibus Appropriations Act, 2009 includes a provision for us to prepare project status reports on selected large-scale NASA programs, projects, and activities.[2] This is our 10th annual report responding to that mandate. This report assesses (1) the cost and schedule performance of NASA's portfolio of major projects, (2) the maturity of critical technologies and stability of project designs at key points in the development process, and (3) the extent to which NASA is assessing future workforce capacity challenges that may affect its ability to manage its portfolio of projects. This report also includes individual assessments of 26 major NASA projects. When NASA determines that a project has an estimated life-cycle cost of over $250 million, we include that project in our annual review up through launch or completion.

To assess the cost and schedule performance, technology maturity, and design stability of NASA's major projects, we collected information on these areas from projects using a data collection instrument, analyzed projects' monthly status reports, interviewed NASA project and headquarters officials, and reviewed project documentation. There are 26 major projects in total, but the information available depends on where a project is in its life cycle.[3] For the 17 projects in the implementation phase, we compared cost and schedule estimates as of February 2018 to their original cost and schedule baselines, identified the number of technologies being developed, and assessed their technology maturity against GAO-identified acquisition best practices and NASA policy. We also compared the number of releasable design drawings at the critical design review against GAO-identified

[1] GAO, *High-Risk Series: Progress on Many High-Risk Areas, While Substantial Efforts Needed on Others*, GAO-17-317 (Washington, D.C.: Feb. 15, 2017).

[2] See Explanatory Statement, 155 Cong. Rec. H1653, 1824-25 (daily ed., Feb. 23, 2009), on H.R. 1105, the Omnibus Appropriations Act, 2009, which became Pub. L. No. 111-8. In this chapter, we refer to these projects as major projects rather than large-scale projects as this is the term used by NASA.

[3] Eight projects were in an early stage of development, called formulation, when there are still unknowns about requirements, technology, and design. For those projects, we reported preliminary cost ranges and schedule estimates. The Commercial Crew Program has a tailored project life cycle and project management requirements. As a result, it was excluded from our cost and schedule performance, technology maturity, and design stability analyses.

acquisition best practices and analyzed subsequent design drawings changes. We reviewed historical data on cost and schedule performance, technology maturity, and design stability for major projects from our prior reports and compared it to the performance of NASA's current portfolio of major projects.

To assess the extent to which NASA is assessing future workforce capacity, we reviewed NASA assessments of technical capabilities and business services, interviewed officials responsible for the assessments and implementing recommendations from the assessments, including officials within the Office of Human Capital Management, and analyzed workforce data found in a NASA database. We determined that the data were sufficiently reliable for our purpose by reviewing responses to data reliability questions from knowledgeable agency officials and the data for obvious errors.

Finally, to conduct our 26 project assessments, we analyzed monthly status reports and interviewed project officials to identify major sources of risk and the strategies that projects are using to mitigate them. Appendix I contains detailed information on our scope and methodology.

We conducted this performance audit from April 2017 to May 2018 in accordance with generally accepted government auditing standards. Those standards require that we plan and perform the audit to obtain sufficient, appropriate evidence to provide a reasonable basis for our findings and conclusions based on our audit objectives. We believe that the evidence obtained provides a reasonable basis for our findings and conclusions based on our audit objectives.

The life cycle for NASA space flight projects consists of two phases— formulation, which takes a project from concept to preliminary design, and implementation, which includes building, launching, and operating the system, among other activities. NASA further divides formulation and implementation into phase A through phase F. Major projects must get approval from senior NASA officials at key decision points before they can enter each new phase. Figure 1 depicts NASA's life cycle for space flight projects.

Project formulation consists of phases A and B, during which the projects develop and define requirements, cost and schedule estimates, and the system's design for implementation. NASA Procedural Requirements 7120.5E, NASA Space Flight Program and Project Management Requirements, specifies that during formulation, the project must complete a formulation agreement to establish the technical and acquisition work that needs to be conducted during this phase and define the schedule and funding requirements for that work. The formulation agreement should identify new technologies and their planned development, the use of heritage technologies, risk mitigation plans, and testing plans to ensure that technologies will work as intended in a relevant environment. Prior to entering phase B, projects develop a range of the project's expected cost and schedule which is used to inform the budget planning for that project. During phase B, the project also develops programmatic measures and technical leading indicators, which track various project metrics such as requirement changes, staffing demands, and mass and power utilization. Near the end of formulation, leading up to the preliminary design review, the project team completes technology development and its preliminary design.

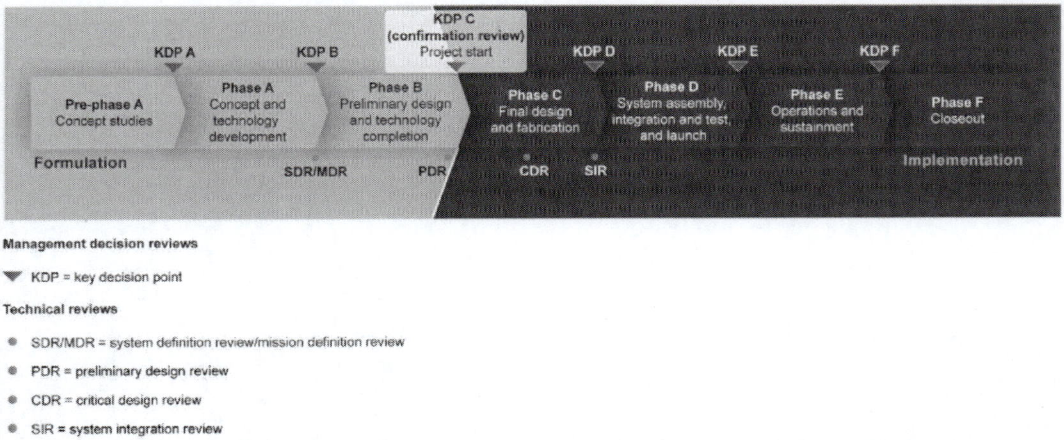

Source: GAO analysis of National Aeronautics and Space Administration data. | GAO-18-280SP.

Figure 1. NASA's Life Cycle for Space Flight Projects.

Formulation culminates in a review at key decision point C, known as project confirmation, where cost and schedule baselines are established and documented in a decision memorandum. The decision memorandum outlines the management agreement and the agency baseline commitment. The management agreement can be viewed as a contract between the agency and the project manager. The project manager has the authority to manage the project within the parameters outlined in the agreement. The agency baseline commitment includes the cost and schedule baselines against which the agency's performance on a project may be measured.

To inform the management agreement and the agency baseline commitment, each project with a life-cycle cost estimated to be greater than $250 million must also develop a joint cost and schedule confidence level (JCL). The JCL initiative, adopted in January 2009, produces a point-in-time estimate that includes, among other things, all cost and schedule elements in phases A through D, incorporates and quantifies known risks, assesses the impacts of cost and schedule to date, and addresses available annual resources. NASA policy requires that projects be baselined and budgeted at the 70 percent confidence level and funded at a level equivalent to at least the 50 percent confidence level.[4]

The management agreement and agency baseline commitment include cost and schedule reserves held at the project and NASA headquarters-level, respectively.[5] Cost reserves are for costs that are expected to be incurred—for instance, to address project

[4] NASA Procedural Requirements (NPR) 7120.5E, *NASA Space Flight Program and Project Management Requirements* paras 2.4.4 and 2.4.4.2 (Aug. 14, 2012) (hereinafter cited as NPR 7120.5E (Aug. 14, 2012)). The decision authority for a project can approve it to move forward at less than the 70 percent confidence level. That decision must be justified and documented.

[5] NASA refers to cost reserves as unallocated future expenses.

risks—but are not yet allocated to a specific part of the project. Schedule reserves are extra time in project schedules that can be allocated to specific activities, elements, and major subsystems to mitigate delays or address unforeseen risks. Project-held cost and schedule reserves are within the project manager's control. If the project requires additional time or money beyond the management agreement—for example, if a project needs additional funds for an issue outside of the project's control—NASA headquarters may allocate headquarters-held reserves. Figure 2 notionally depicts how NASA would distribute cost reserves for a project that was baselined in accordance with its JCL policy.

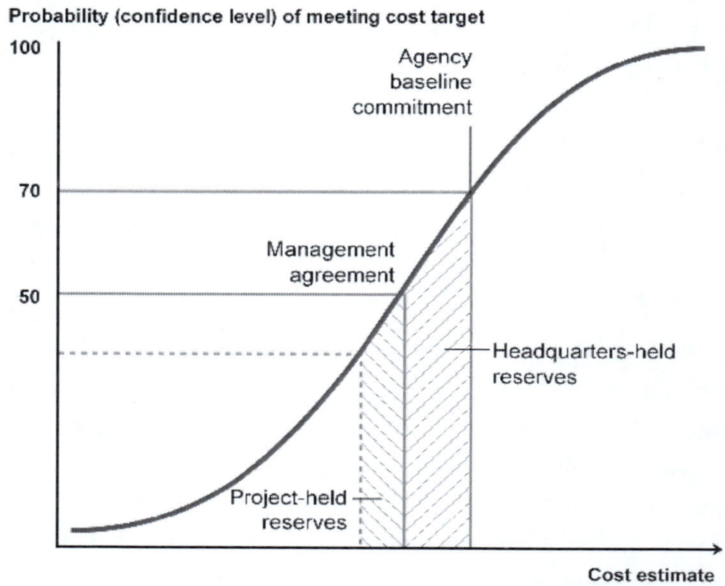

Source: GAO analysis of National Aeronautics and Space Administration data. | GAO-18-280SP.

Figure 2. Notional Distribution of Cost Reserves for a Project Budgeted at the 70 Percent Confidence Level.

The total amount of cost and schedule reserves held at the project level varies based on where the project is in its life cycle. Seven NASA centers or laboratories are responsible for managing 25 NASA major projects. Of these, four centers or laboratories manage 20 of the 25 major projects and require or recommend that projects hold a certain level of cost and schedule reserves at key project milestones.[6] For example, at the Goddard Space Flight

[6] Goddard Procedural Requirements 7120.7A, *Schedule and Budget Margins for Flight Projects* (Feb. 28, 2017); Marshall Procedural Requirements 7120.1, *Marshall Space Flight Center Engineering and Program/Project Management Requirements* (Aug. 26, 2014); Langley Research Center, *Space Flight Project Practices Handbook*, LPR 7120.5 B-1 (Mar. 17, 2014); and Jet Propulsion Laboratory, *Flight Project Practices, Rev. 8* (Oct. 6, 2010). The Kennedy Space Center and Johnson Space Center do not have center-specific guidance for reserves. The Johns Hopkins University Applied Physics Laboratory manages the Parker Solar Probe (PSP) and Double Asteroid Redirection Test (DART) projects and has guidelines for schedule reserves, but not for cost reserves. The Johns Hopkins University Applied Physics Laboratory SD-QP-012, Rev. b, *Space Exploration Sector (SES) Quality Procedure: Earned Value Management System (EVMS) Project Management Control System (PMCS)* (Apr. 4, 2017).

Center, projects are required to hold cost reserves equal to at least 25 percent of the estimated cost remaining at the project confirmation review, and 10 percent at the time of delivery to the launch site. Projects track their reserves between phases to help ensure they hold reserves consistent with these requirements. The 26th major project included in our review, the Low Boom Flight Demonstrator (LBFD), does not have a lead center because it is using a virtual project office model. Project officials said they have not yet determined which center cost and schedule guidelines the project will follow.

After a project is confirmed, it begins implementation, consisting of phases C, D, E, and F. In this report, we refer to projects in phase C and D as being in development. A critical design review is held during the latter half of phase C in order to determine if the design is mature enough to support proceeding with the final design and fabrication. After the critical design review and just prior to beginning phase D, the project completes a system integration review to evaluate the readiness of the project and associated supporting infrastructure to begin system assembly, integration and test. In phase D, the project performs system assembly, integration, test, and launch activities. Phases E and F consist of operations and sustainment and project closeout.

NASA Projects Reviewed in GAO's Annual Assessment

NASA's portfolio of major projects ranges from satellites equipped with advanced sensors to study the Earth to a rover that plans to collect soil or rock samples on Mars to telescopes intended to explore the universe to spacecraft to transport humans and cargo beyond low-Earth orbit. When NASA determines that a project will have an estimated life-cycle cost of more than $250 million, we include that project in our annual review. After a project launches or reaches full operational capability, we no longer include an assessment of it in our annual report.

This chapter includes assessments of 26 major NASA projects (see Table 1). Six projects are being assessed for the first time this year: Double Asteroid Redirection Test (DART), Laser Communications Relay Demonstration (LCRD), LBFD, Lucy, Psyche, and Restore-L. Prior to establishing its cost baseline, the LCRD project had an estimated life-cycle cost estimate of less than $250 million and therefore was not included in our previous reviews when the project was in formulation. For a list of the 26 projects and their current cost and schedule estimates, see appendix II. We did not complete an assessment for the Asteroid Redirect Robotic Mission project this year because, as of June 2017, the project closed out most project development activities at the direction of NASA headquarters. Appendix III includes a list of all the projects that we have reviewed from 2009 to 2017.

Table 1. Major NASA Projects Reviewed in GAO's 2018 Assessment by Phase

Projects in formulation	Double Asteroid Redirection Test (DART) Europa Clipper
	Low Boom Flight Demonstrator (LBFD) Lucy
	Plankton, Aerosol, Cloud, ocean Ecosystem (PACE) Psyche Restore-L
	Wide-Field Infrared Survey Telescope (WFIRST)
Projects in implementation	Commercial Crew Program (CCP)
	Exploration Ground Systems (EGS)
	Gravity Recovery and Climate Experiment Follow-On (GRACE-FO)
	Ice, Cloud, and Land Elevation Satellite-2 (ICESat-2)
	Interior Exploration using Seismic Investigations, Geodesy, and Heat Transport (InSight)
	Ionospheric Connection Explorer (ICON)
	James Webb Space Telescope (JWST)
	Landsat 9 (L9)
	Laser Communications Relay Demonstration (LCRD)
	Mars 2020
	NASA ISRO Synthetic Aperture Radar (NISAR)
	Orion Multi-Purpose Crew Vehicle (Orion)
	Parker Solar Probe (PSP) (formerly Solar Probe Plus)
	Radiation Budget Instrument (RBI)
	Space Launch System (SLS)
	Space Network Ground Segment Sustainment (SGSS)
	Surface Water and Ocean Topography (SWOT)
	Transiting Exoplanet Survey Satellite (TESS)

Source: GAO analysis of National Aeronautics and Space Administration data. | GAO-18-280SP.

Over the past 6 years, we have issued several reports assessing NASA's progress acquiring its largest projects and programs in more depth.[7] We found in July 2016 that all three human spaceflight programs—the Orion Multi-Purpose Crew Vehicle (Orion), the

---

[7] GAO, *James Webb Space Telescope: Actions Needed to Improve Cost Estimate and Oversight of Test and Integration*, GAO-13-4 (Washington, D.C.: Dec. 3, 2012); *Space Launch System: Resources Need to be Matched to Requirements to Decrease Risk and Support Long Term Affordability*, GAO-14-631 (Washington, D.C.: July 23, 2014); *James Webb Space Telescope: Project Meeting Commitments but Current Technical, Cost, and Schedule Challenges Could Affect Continued Progress*, GAO-14-72 (Washington, D.C.: Jan. 8, 2014); *James Webb Space Telescope: Project Facing Increased Schedule Risk with Significant Work Remaining,* GAO-15-100 (Washington, D.C.: Dec. 15, 2014); *Space Launch System: Management Tools Should Better Track to Cost and Schedule Commitments to Adequately Monitor Increasing Risk,* GAO-15-596 (Washington, D.C.: July 16, 2015); *James Webb Space Telescope: Project on Track but May Benefit from Improved Contractor Data to Better Understand Costs,* GAO-16-112 (Washington, D.C.: Dec. 17, 2015);*NASA Human Space Exploration: Opportunity Nears to Reassess Launch Vehicle and Ground Systems Cost and Schedule,* GAO-16-612 (Washington, D.C.: July 27, 2016); *Orion Multi-Purpose Crew Vehicle: Action Needed to Improve Visibility into Cost, Schedule, and Capacity to Resolve Technical Challenges,* GAO-16-620 (Washington, D.C.: July 27, 2016); *NASA Commercial Crew Program: Schedule Pressure Increases as Contractors Delay Key Events,* GAO-17-137 (Washington, D.C.: Feb. 16, 2017); *NASA Human Space Exploration: Delay Likely for First Exploration Mission,* GAO-17-414 (Washington, D.C.: Apr. 27, 2017); *NASA Human Space Exploration: Integration Approach Presents Challenges to Oversight and Independence,* GAO-18-28 (Washington, D.C.: Oct 19, 2017); *NASA Commercial Crew Program: Continued Delays Pose Risks for Uninterrupted Access to the International Space Station,* GAO-18-317T (Washington, D.C.: Jan. 17, 2018); and *James Webb Space Telescope: Integration and Test Challenges Have Delayed Launch and Threated to Push Costs Over Cap,* GAO-18-273 (Washington, D.C.: Feb. 28, 2018).

Space Launch System (SLS), and Exploration Ground Systems (EGS)—were making progress in resolving technical issues and maturing designs, but that pressure on the limited cost and schedule reserves put the schedule for their first combined mission, the uncrewed Exploration Mission-1 (EM-1), at risk.8 Subsequently, in April 2017, we found that given the combined effects of ongoing technical challenges in conjunction with limited cost and schedule reserves, it was unlikely that these programs would achieve the planned November 2018 launch readiness date.9 We recommended that NASA confirm whether this launch readiness date was achievable and, if warranted, propose a new, more realistic EM-1 date and report to Congress on the results of its schedule analysis. NASA agreed with both recommendations and stated that it was no longer in its best interest to pursue the November 2018 launch readiness date. In December 2017, NASA approved a new EM-1 schedule of December 2019, with six months of schedule reserve available to extend the date to June 2020. In April 2018, NASA provided us with the letter the agency sent to Congress on its assessment of the EM-1 schedule. We are in the process of reviewing the letter to determine if it addresses our recommendation.

We have also reported for several years on the James Webb Space Telescope (JWST) project, which has experienced significant cost increases and schedule delays. Prior to being approved for development, cost estimates for JWST ranged from $1 billion to $3.5 billion, with expected launch dates ranging from 2007 to 2011. Before 2011, early technical and management challenges, contractor performance issues, low levels of cost reserves, and poorly phased funding levels caused JWST to delay work after confirmation, which contributed to significant cost and schedule overruns, including launch delays. Following an independent review that found JWST was executing well from a technical standpoint, but that the baseline cost estimate did not reflect the most probable cost with adequate reserves in each year of project execution, Congress placed an $8 billion cap on the formulation and development costs for the project in November 2011. NASA rebaselined JWST with a life-cycle cost estimate of $8.835 billion that included additional money for operations and a planned launch in October 2018.[10] In December 2017, we found that the JWST project continues to make progress towards launch, but the program is encountering technical challenges that require both time and money to fix and may lead to additional delays.[11]

---

[8] GAO-16-620 and GAO-16-612.

[9] GAO-17-414.

[10] A rebaseline is a process initiated if development costs increase by 30 percent or more and requires the NASA Administrator to transmit a report to the Committee on Science, Space, and Technology of the House of Representatives and the Committee on Commerce, Science, and Transportation of the Senate.

[11] GAO, *NASA: Preliminary Observations on the Management of Space Telescopes*, GAO-18-277T (Washington, D.C.: Dec. 6, 2017).

NASA's Workforce

In January 2018, NASA had a civil servant workforce of approximately 17,100 employees distributed among NASA Headquarters in Washington, D.C., and nine centers located around the country. Figure 3 shows the distribution of this workforce across NASA headquarters and the centers.

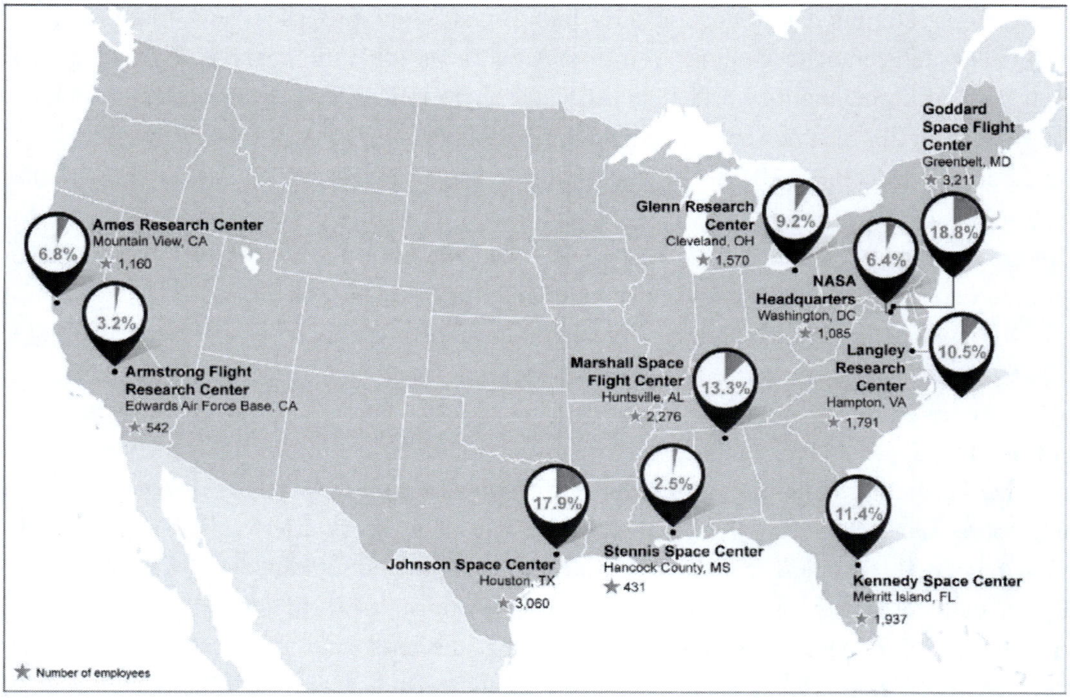

Source: GAO analysis of National Aeronautics and Space Administration data. | GAO-18-280SP.

Note: Civil servant staff within NASA's Shared Services Center are included in the total for the Stennis Space Center because the center is located there.

Figure 3. NASA's Workforce Distribution at NASA Headquarters and Centers as of January 2018.

NASA also relies on a contractor workforce to accomplish its mission. NASA estimated that approximately 33,100 contractors would support NASA in 2018. The Jet Propulsion Laboratory is a federally funded research and development center under contract with NASA and is managed by the California Institute of Technology. It has about 5,500 employees, who primarily conduct NASA-related work. The Applied Physics Laboratory is a research center affiliated with Johns Hopkins University and has about 6,300 employees, some of whom manage and support NASA projects.

Portfolio Cost and Schedule Deteriorated but Extent of Cost Growth Is Unknown

The cost and schedule performance of NASA's portfolio of major projects has deteriorated since last year, but the extent of cost performance deterioration is unknown. NASA lacks a current cost estimate for its Orion crew capsule—one of the largest programs in the portfolio—but expects the program will exceed its cost baseline when NASA updates the program's life-cycle cost estimate. Because the Orion program accounts for about 22 percent of all development costs, even a small percentage of cost growth for the Orion program could significantly affect cost performance. The known negative cost and schedule performance is largely driven by the cost and schedule growth of four projects—SLS, EGS, Space Network Ground Segment Sustainment (SGSS) and Mars 2020—that experienced technical problems compounded by programmatic challenges. Together, these projects experienced $638 million in cost growth and 59 months in aggregate schedule delays. Two projects—JWST and ICESat-2—experienced schedule delays due to technical challenges identified during integration and test. Another 3 projects—NASA ISRO Synthetic Aperture Radar (NISAR), Ionospheric Connection Explorer (ICON), and Gravity Recovery and Climate Experiment Follow-On (GRACE-FO)—experienced cost growth or delays largely due to factors outside of the projects' control, such as launch vehicle delays.

NASA is at risk of continued cost increases and schedule delays as newer, complex projects begin development and expensive projects linger in the portfolio longer than expected due to delays. Several projects that experienced schedule delays or cost increases will face further challenges completing integration and test—the phase of development that often reveals unforeseen challenges that can lead to cost and schedule growth—that could result in additional cost and schedule growth.

Portfolio Average Launch Delays Increased, but NASA Lacks a Current Orion Program Cost Estimate to Determine Extent of Cost Growth

Since we last reported in May 2017, the average launch delay increased to 12 months, up from 7 months in 2017—the highest schedule delay we have reported to date.[12] We are not able to determine the extent of portfolio cost growth this year because NASA does not have a current cost estimate for the Orion program—one of the largest programs in its portfolio—and officials expect the cost to increase. As of June 2017, the Orion program's development cost was about $6.6 billion and based on that estimate it accounts for 22

[12] GAO, *NASA: Assessments of Major Projects*, GAO-17-303SP (Washington, D.C: May 16, 2017).

percent of the portfolio's estimated $30.1 billion of development costs. As a result, a small percentage of cost growth for the Orion program could significantly affect cost performance. Even without including Orion cost growth, the overall development cost growth for the portfolio of 17 development projects increased to 18.8 percent, up from 15.6 percent in 2017 (see Figure 4).[13]

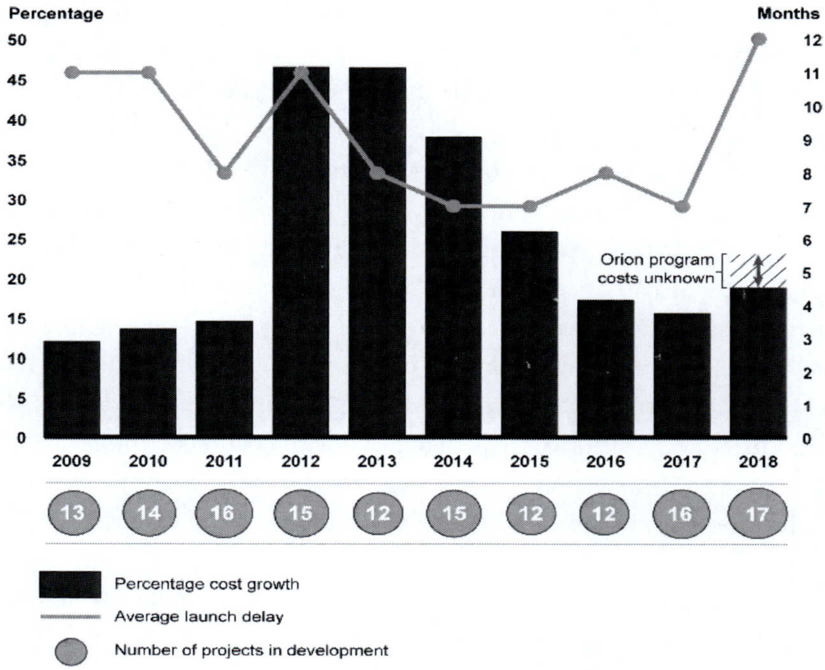

Source: GAO analysis of National Aeronautics and Space Administration data. | GAO-18-280SP.

Note: The years in the figure are the year we issued our report. Cost and schedule performance is compared across each report period.

Figure 4. Development Cost Performance and Average Launch Delay for Major NASA Projects from 2009 to 2018.

Senior-level NASA officials told us that they expect that the Human Exploration and Operations Mission Directorate and the Orion program will complete an updated life-cycle cost estimate in June 2018. This would be approximately 10 months after the program raised to senior-level officials' attention that the program expects cost growth over its cost baseline to the second combined mission, Exploration Mission 2 (EM-2), during an August 2017 briefing concerning potential cost increases related to the EM-1 launch delay.

[13] We have historically presented cost and schedule performance including and excluding the James Webb Space Telescope (JWST) because, prior to 2015, it had a development cost baseline significantly larger than other projects and the magnitude of its cost growth masked the performance of the remainder of the portfolio. Now that other projects in the portfolio, such as Orion and the Space Launch System, have large development cost baselines, we no longer present cost performance trends excluding JWST.

Table 2. Development Cost and Schedule Performance of Selected Major NASA Projects Currently in Development

Overall performance	Project	Confirmation date	Changes since May 2017		Cumulative performance	
		Year	Cost (millions)	Schedule (months)	Cost (millions)	Schedule (months)
Lower than expected cost	PSP	2014	$0.0	0	-$5.4	0
	GRACE-FO	2014	-$1.6	3	-$2.2	3
	TESS	2014	-$13.1	-2	-$39.9	-2
Within baseline	ICON	2014	$0.0	8	$0.0	8
	SWOT	2016	$0.0	0	$0.0	0
	LCRD	2017	$0.0	0	$0.0	0
	Landsat 9	2017	$0.0	0	$0.0	0
Higher than expected cost	Mars 2020[a]	2016	$12.9	0	$10.7	0
	NISAR	2016	$22.0	0	$22.0	0
Replan[c]	InSight	2014	$0.0	0	$131.7	26
	EGS (EM-1)	2014	$417.8	19	$421.4	19
	SLS (EM-1)	2014	$147.8	19	$147.8	19
Rebaseline[c]	JWST	2008	$0.0	19	$3,607.7	71
	ICESat-2	2012	$1.4	4	$206.3	17
	SGSS[b]	2013	$59.5	21	$421.6	48
Canceled	RBI	2016	$0.0	0	$0.0	0
Under revision	Orion (EM-2)[d]	2015	$0.0	0	-$151.7	0
Total:			$646.7	91	$4,770.0	209

Legend: PSP: Parker Solar Probe; GRACE-FO: Gravity Recovery and Climate Experiment Follow-On; TESS: Transiting Exoplanet Survey Satellite; ICON: Ionospheric Connection Explorer; SWOT: Surface Water and Ocean Topography; NISAR: NASA Indian Space Research Organisation – Synthetic Aperture Radar; LCRD: Laser Communications Relay Demonstration; InSight: Interior Exploration using Seismic Investigations, Geodesy, and Heat Transport; EGS: Exploration Ground Systems; SLS: Space Launch System; EM-1: Exploration Mission 1; JWST: James Webb Space Telescope; ICESat-2: Ice, Cloud, and Land Elevation Satellite-2; SGSS: Space Network Ground Segment Sustainment; RBI: Radiation Budget Instrument; Orion: Orion Multi-Purpose Crew Vehicle; EM-2: Exploration Mission 2.

Source: GAO analysis of National Aeronautics and Space Administration data. | GAO-18-280SP.

Note: Positive values indicate cost growth or launch delays. Negative values indicate cost decreases or earlier than planned launch dates.

[a]The Mars 2020 project used $2.2 million in funds originally budgeted for development for formulation activities. This partially offsets an increase of $12.9 million in development cost growth primarily due to increased costs associated with a technology demonstration instrument and entry, descent, and landing instrument.

[b]The SGSS project reported cost growth up through its first operational readiness review, which is currently planned for the end of fiscal year 2019. However, the project expects that there could be additional cost and schedule growth beyond what is reported here.

[c]A replan is a process initiated if development costs increase by 15 percent or more. NASA replanned the SLS program even though development costs did not increase by 15 percent or more. A replan does not require a new project baseline to be established. A rebaseline is a process initiated if development costs increase by 30 percent or more. Both processes require NASA to submit a report to relevant congressional committees. In addition, if a project or program milestone is likely to be delayed by 6 months or more, this report is also required.

[d]NASA officials said they are revising the Orion program's life-cycle cost estimate and expect to complete a new estimate in June 2018. The new cost is expected to exceed the program's development cost baseline. The current costs in the table reflect the estimate provided in June 2017. The cumulative cost change reflects the program shifting $151.7 million of funding previously budgeted for the development phase to the formulation phase.

Seven of 17 NASA major projects stayed within cost and schedule estimates since we last reported in May 2017, but 9 projects experienced cost growth or schedule delays and cost growth is expected for the Orion program. Table 2 provides data on the cost and schedule performance for the 17 major projects in development that have cost and schedule baselines since we last reported on the portfolio's performance in May 2017.[14]

The deteriorating cost and schedule performance of the portfolio in 2018 is the result of

- four projects—SLS, EGS, SGSS, and Mars 2020—addressing technical challenges that were compounded by risky programmatic decisions;
- two projects—JWST and ICESat-2—experiencing delays due to technical challenges identified during integration and test; and
- three projects—NISAR, ICON, and GRACE-FO—experiencing cost growth or delays largely due to factors outside of the projects' control.

We elaborate on these three scenarios below.

Technical Challenges Compounded by Risky Programmatic Decisions

Together, SLS, EGS, SGSS, and Mars 2020 experienced $638 million in cost growth and 59 months in aggregate schedule delays due to technical problems that were compounded by programmatic challenges since our May 2017 report.[15] The SLS and EGS programs experienced cost growth and schedule delays associated with EM-1, their first combined mission along with the Orion program. We have found for several years that the human spaceflight programs—Orion, SLS, and EGS—are making progress maturing designs and building hardware, but also are experiencing some significant engineering and manufacturing challenges. For example, the SLS program ran into numerous challenges completing the welding of its core stage element in 2017. The program stopped welding on the core stage for months to identify and resolve low weld strength in the liquid oxygen and liquid hydrogen tanks due to low weld strength measurements found in the liquid oxygen tanks caused by a program and contractor decision to change the weld tool configuration during fabrication. The EGS program also experienced technical challenges, including with the design and installation of the ground support equipment and the 10 umbilicals that connect SLS and Orion to the Mobile Launcher—which supports the assembly, testing, and servicing of SLS and provides the platform on which SLS and Orion will launch.

Finally, although the Orion program has not yet reported cost growth, it also experienced technical challenges. These challenges included software delays and hardware

[14] GAO-17-303SP.
[15] GAO-17-303SP.

delays, and at least 14 months of delays with the European Service Module—which provides air, water, power, and propulsion to Orion during in-space flight—since the element's critical design review in June 2016. In April 2017, we found that, according to program officials, the delays with the service module were largely due to NASA, the European Space Agency, and the European Space Agency contractor underestimating the time and effort necessary to address design issues for the first production service module and the availability of parts from suppliers and subcontractors. NASA expects the Orion program to experience cost growth over its cost baseline to the second combined mission, EM-2. However, the extent of the growth is unknown because, as noted above, NASA is currently revising its life-cycle cost estimate.

Technical challenges such as these are not unusual for large-scale programs, especially human exploration programs that are inherently complex and difficult. However, we have found that NASA has made programmatic decisions—including establishing low cost and schedule reserves, managing to aggressive schedules, and not following best practices for earned value management or creating reliable cost and schedule baselines—that have compounded the technical challenges (see Table 3). As a result, the three human spaceflight programs have been at risk of cost and schedule growth since NASA approved their baselines.

In December 2017, NASA announced the new internal launch readiness date for EM-1 is now December 2019, and has allocated 6 months of schedule reserve available to extend the date to June 2020 for possible manufacturing and production schedule risks. This represents a delay of 13-19 months for EM-1. It is too soon to know if NASA has addressed the programmatic challenges identified above but we will continue to follow up through future reviews.

Similarly, the SGSS project experienced new cost growth of $59.5 million and delayed its completion by 21 months. Project officials attributed the cost growth and delays to the contractor's incomplete understanding of its requirements, which led to poor contractor plans and late design changes. But project management has been a challenge as well.[16] The project has historically struggled to manage contractor performance and has faced both contractor and project staffing shortfalls, as we found in our prior reports starting in 2013.[17] For example, NASA managers noted concerns with contractor plans and staffing estimates in 2013 during project confirmation. In March 2015, we found that the project was being

[16] In 2016, NASA announced it was reclassifying SGSS as a hybrid sustainment project for the Space Network. A hybrid sustainment effort is a sustainment effort that still includes development work. The SGSS project expects to experience additional cost growth and schedule delays, but the exact magnitude is unknown. The project was reevaluating its cost and schedules through its final acceptance review at the time of our review.

[17] GAO, *NASA: Assessments of Selected Large-Scale Projects*, GAO-13-276SP (Washington, D.C.: Apr. 17, 2013); *NASA: Assessments of Selected Large-Scale Projects*, GAO-14-338SP (Washington, D.C.: Apr. 15, 2014); *NASA: Assessments of Selected Large-Scale Projects*, GAO-15-320SP (Washington, D.C.: Mar. 24, 2015); *NASA: Assessments of Major Projects,* GAO-16-309SP (Washington, D.C.: Mar. 30, 2016); and GAO-17-303SP.

Table 3. Examples of Risky Programmatic Decisions Made by NASA's Human Spaceflight Programs

Programmatic decision	Example(s)	Negative effect	GAO report(s)
NASA baselined the Exploration Ground Systems (EGS), Orion Multi-Purpose Crew Vehicle (Orion), and Space Launch System (SLS) programs with low cost and schedule reserves.	In July 2016, we found that NASA baselined the SLS program with cost reserves of less than 2 percent, even though guidance for Marshall Space Flight Center—the NASA center with responsibility for the SLS program—established standard cost reserve for launch vehicle programs of 20 percent when the baseline is approved.	Operating with low cost and schedule reserves limits a program's ability to address risks and unforeseen technical challenges.	GAO-17-414 and GAO-16-612
NASA managed the EGS, Orion, and SLS programs to an internal schedule for completing development production that was aggressive and could exacerbate delays and lead to cost overruns.	In July 2016, we found that the EGS program planned to conduct the mobile launcher's verification and validation concurrent with ground support equipment systems and umbilicals installation to support the program's internal schedule goal. We found this to be a risky practice because of uncertainties regarding how systems not yet installed may affect the systems already installed.	Working towards a more aggressive internal goal is not a bad practice; however, increasing cost and schedule risk to the program in order to pursue such a goal is not beneficial to programs in the long term.	GAO-16-620 and GAO-16-612
The SLS program did not follow best practices for using earned value management, which integrates the project scope of work with cost, schedule, and performance elements for optimum project planning and control.	In July 2016, we found that the SLS program had not positioned itself well to provide accurate assessments of progress with the core stage because it operated for several years without a performance measurement baseline that is necessary to support full earned value management reporting. The use of earned value management is advocated by both GAO's best practices for cost estimating and NASA's own guidance.	Programs that do not use earned value data are limited in their ability to have accurate assessments of project progress, produce early warning signs of impending schedule delays and cost overruns, and provide unbiased estimates of anticipated costs at completion.	GAO-16-612
The Orion and SLS programs established baselines that were not fully reliable.[a]	In July 2016, we found that the Orion program did not generally follow best practices in preparing its cost and schedule estimates, which were key inputs into the program's joint cost and schedule confidence level processes and baseline. In July 2015, we found that cost and schedule estimates for the SLS program substantially met five of six characteristics that GAO considers best practices for preparing reliable estimates, but could not be deemed fully reliable because they only partially met the sixth characteristic— credibility.	Without sound cost and schedule estimates, decision makers do not have a clear understanding of the cost and schedule risk inherent in the program or important information needed to make programmatic decisions.	GAO-16-620 and GAO-15-596

Source: GAO analysis of prior GAO reports. | GAO-18-280SP.

[a]We did not assess EGS's cost and schedule estimates compared to best practices.

rebaselined due to the contractor's poor cost and schedule performance and limitations that NASA placed on the funding available to the contractor in fiscal years 2014 and 2015. The contractor was also operating with a limited number of staff at that time. In May 2017, we found that the project continued to experience contractor performance problems and had experienced cost growth and schedule delays over the 2015 rebaseline even as the project decreased its scope. In addition, the project experienced staff shortfalls in key areas, such as systems engineering and business management.

The Mars 2020 project experienced $12.9 million in development cost growth, but no schedule delays. The cost growth was primarily due to technical challenges on a technology demonstration instrument and higher than anticipated integration costs for an entry, descent, and landing instrument. Both instruments are funded by the Human Exploration and Operations and Space Technology Mission Directorates. NASA officials attributed the cost growth of the technology demonstration instrument—which is designed to convert carbon dioxide to oxygen—to the complexity of the technology development for the effort. At the project's preliminary design review in February 2016, a critical technology for the technology demonstration instrument did not meet the recommended level of maturity, which we have found can increase risk for systems entering product development. The project had matured the technology to this recommended level by its critical design review in February 2017. However, as a result of the focus on maturing this particular technology, other components of the instrument fell behind the planned schedule. Project costs for Mars 2020 also increased for an entry, descent, and landing instrument, due, in part, to cost increases for integration and to add additional staff to the instrument team to maintain schedule.

Finally, the RBI project would have likely exceeded its cost baseline if NASA had not decided to cancel the project in January 2018. According to NASA's cancellation memorandum, the project was canceled because of continued cost growth, technical issues, and poor contractor performance. In 2017, we found that the project was working to an aggressive schedule, and the prime contractor continued to experience cost overruns even after NASA added a deputy project manager and increased site visits and meetings with the contractor.[18] Subsequently, the project—which was developing an instrument to be hosted on a National Oceanic and Atmospheric Administration satellite—determined that it would not be able to meet its delivery date for integration with the satellite without requiring additional funding in excess of the project's cost baseline if other technical issues arose. In its cancellation memorandum, NASA stated continuing to fund RBI from within the Earth Science Division budget would slow other important activities.

[18] GAO-17-303SP.

Technical Challenges Identified during Integration and Test

The JWST and ICESat-2 projects experienced technical challenges during integration and test that delayed their schedules. Both projects were previously rebaselined before entering system-level integration and testing, and the current schedule delays are beyond the new schedules that NASA set for the projects in 2011 for JWST and in 2014 for ICESat-2.

- The JWST project delayed its launch readiness date by at least 19 months from October 2018 to May 2020. NASA announced two delays for the project since our last portfolio-wide review in May 2017.[19] First, as we found in February 2018, the project delayed its launch readiness date by up to 8 months primarily due to the integration of the various spacecraft elements taking longer than expected.[20] Specifically, execution of spacecraft integration and test tasks, due to complexity of work and cautious handling given the sensitivity of flight hardware, was slower than planned. In addition, before the delay, the project used all of its schedule reserves to its prior launch readiness date. This was the result of various contractor workmanship errors, particularly with respect to the spacecraft propulsion systems, as well as the resolution of various technical issues, including a test anomaly on the telescope and sunshield hardware challenges. Second, in March 2018, NASA announced that it had delayed the project's launch readiness date by an additional 11 months to approximately May 2020 and planned to establish an external independent review board to analyze the project's organizational and technical issues to inform a more specific launch time frame.
- The announcement also stated that after a new launch date is established, NASA would provide a new cost estimate that may exceed the $8 billion congressional cost cap that was established in 2011. NASA plans to finalize the project's cost and schedule in June 2018. Because the additional delays were announced while a draft of this report was with NASA for comment, we plan to follow up on the reasons for the additional delays and the results of the analysis in a future review.
- The ICESat-2 project delayed its launch readiness date by 4 months from June to October 2018 due to technical issues with its only instrument, the Advanced Topographic Laser Altimeter System. A key part in the instrument's lasers failed during instrument environmental testing, which delayed the project's system integration review—the start of system-level integration and test. The manufacturer determined the primary cause of the anomaly was a flaw in the design of the mount that ensures a component of the optical module remains in a specific, precise position. The spare flight laser encountered the same problem

[19] GAO-17-303SP.
[20] GAO-18-273.

during earlier testing, which indicated a systemic problem. The project redesigned and repaired the lasers and is proceeding through integration and test.

External Factors

External factors—including responding to requests for additional data collection and delays due to launch-vehicle related issues—contributed to cost increases or schedule delays for the NISAR, ICON, and GRACE-FO projects.

- The NISAR project experienced cost growth as the result of an increase in the scope of data collection in response to additional data needs being identified by an interagency working group. The additional data includes soil moisture and natural hazard data that would be of value for other federal agencies and the science community. NASA officials said the additional funding for development would be used to upgrade the ground stations so that they can receive the additional data at a higher downlink data rate and volume.
- The ICON project missed its committed launch readiness date because of an accident involving its launch vehicle. In January 2017, two of the Pegasus launch vehicle's three stages were involved in a transport accident. The stages were subsequently returned to the launch vehicle contractor facility for inspection and testing, and no damage was found. The project had been on track to launch early. Subsequently, in September 2017, an anomaly found in testing of the launch vehicle bolt cutter assemblies resulted in additional delays, but the magnitude of these delays is not yet known; an investigation is underway.
- The GRACE-FO project delayed its launch readiness date from February to May 2018 due to issues with its planned launch vehicle and launch site. The launch vehicle is the responsibility of NASA's partner on the project—German Research Centre for Geosciences (GFZ). GRACE-FO had planned to launch at a Russian launch site. In February 2016, GFZ reported that it was notified by the Russian Federal Space Agency that the Dnepr launch vehicle was no longer available for GRACE-FO. GFZ, in June 2016, arranged to launch the two GRACE-FO spacecraft, along with commercial satellites, on a SpaceX Falcon 9.

In addition, the Commercial Crew Program also experienced delays, which are not included above because the program does not have a schedule baseline. Since the award of the current Commercial Crew contracts in September 2014, the program, Boeing, SpaceX, and multiple independent review bodies have all identified the contractors' delivery schedules as aggressive. In February 2017, we found that Boeing and SpaceX had determined that neither could meet their original 2017 dates for NASA to certify their systems for human spaceflight.[21] In January 2018, we found that both contractors had notified NASA that final certification dates have slipped again and are now in the first quarter of calendar year 2019.[22] The Commercial Crew Program's schedule analysis indicates that certification may be further delayed to December 2019 for SpaceX and February 2020 for Boeing.

NASA Is at Risk of Continued Cost Increases and Schedule Delays as Newer, Complex Projects Begin Development and Expensive Projects Linger

In 2016 and 2017, we found that projects appear most likely to rebaseline between their critical design review and system integration review—the riskiest point in the development cycle.[23] Of the 14 projects in NASA's current portfolio of major projects that have held a critical design review, 3 had to rebaseline before holding a system integration review. An additional 4 projects experienced cost growth after critical design review, but the cost growth did not require a rebaseline. Further, most of the projects that have experienced cost growth during this risky point in the development cycle are not small projects. All but one has a current life-cycle cost estimate of over $1 billion and include NASA's flagship missions, like JWST and Mars 2020. The total development cost growth for these 7 projects is about $4.9 billion to date, but additional cost growth could occur as the projects are not yet complete.

The composition of the portfolio in the coming years is expected to include similarly large and complex projects, putting NASA at risk of continued cost increases and schedule delays. Specifically, NASA plans to have complex projects enter the development portfolio in the next few years as they hold confirmation reviews and set cost and schedule baselines. This includes the Europa Clipper project and potentially the Wide-Field Infrared Survey Telescope (WFIRST) project. In February 2018, the President's 2019 Budget Request proposed canceling the WFIRST project due to the project's significant costs and higher priorities in the agency. However, the project may continue if funding is received.

Together, preliminary estimates indicate that these two projects could cost as much as $7.8 billion. In addition, NASA expects to begin other large, complex projects like the Lunar Orbital Platform-Gateway— currently being discussed as a space station or outpost in lunar orbit— and a Europa Lander project in the coming years. A December 2017 space policy directive also instructed NASA to return astronauts to the moon for long-term exploration and to pursue human exploration of Mars and the broader solar system.

To its credit, NASA recently took steps to put a process in place to control the costs of two projects while in formulation, which may prove useful if properly executed.

---

[21] GAO-17-137.
[22] GAO-18-317T.
[23] GAO-16-309SP and GAO-17-303SP.

- The Europa Clipper project implemented a process whereby cost growth threats would be offset by descoping instruments in whole or in part. For example, if an instrument exceeds its development cost by 20 percent, the project would propose a descope option to NASA that brings instrument cost below that threshold. NASA has not descoped any instruments to date.
- The WFIRST project is responding to findings from an independent review that was conducted to ensure the mission's scope and required resources are well understood and executable. The review found that the mission scope is understood, but not aligned with the resources provided and concluded that the mission is not executable without adjustments and/or additional resources. For example, the study team found that NASA's current forecasted funding profile for the WFIRST project would require the project to slow down activities starting in fiscal year 2020, which would result in an increase in development cost and schedule. NASA agreed with the study team's results and directed the project to reduce the cost and complexity of the design in order to maintain costs within the $3.2 billion preliminary cost target.

But even with these efforts, NASA's cost and schedule performance may be further tested in upcoming years as some expensive, complex projects linger in the portfolio longer than expected.

- As previously discussed, the Orion program expects cost growth and faces other schedule and technical risks as it moves through the integration and test phase for EM-1 into at least 2019 and then through 2023 for EM-2. As of August 2017, NASA officials expected that new hardware and addressing development challenges would be the factors contributing to increased cost for the program. For example, there was a cost impact when the program moved from a single-piece, or monolithic, heatshield design to one that employs blocks in order to improve its structural strength. Program officials said they are also assessing schedule delays for EM-2, and noted that the EM-2 launch date depends on the outcome of the EM-1 launch date.
- The SLS and EGS programs continue to face cost, schedule, and technical risks as they move through the integration and test phase into at least 2019. For example, SLS will have to complete a "green run" test which requires multiple first-time efforts. Specifically, the test is the culmination of the development effort and includes the core stage integration with its four main engines, fully fueling with cryogenic hydrogen and oxygen, and then firing all four engines for about 500 seconds. NASA currently has no schedule reserve to its target December 2019 launch readiness date for two key areas in the core stage schedule. First, there is no reserve between the end of core stage production and the delivery of the core

stage to the test facility. Second, there is no reserve between the end of the testing and delivery to Kennedy Space Center for final integration and testing prior to launch.

- As previously discussed, the JWST project is at risk of exceeding its congressional cost cap, and faces schedule risks as it completes its remaining integration and test work. These activities have taken considerably longer than planned due to a variety of challenges, including reach and access limitations on the flight hardware. Additionally, the project faces significant work ahead. For example, the project must complete integration of spacecraft element hardware and conduct deployment and environmental tests of the integrated sunshield and spacecraft. Further, it must integrate the telescope element with the spacecraft element to form the JWST observatory, and complete another set of challenging environmental tests on the full integrated observatory. At the same time, the project will need to mitigate dozens of remaining hardware and software risks to acceptable levels and address the project's many potential single point failures to the extent possible.
- The SGSS project expects to experience additional cost growth through the final acceptance review because the full scope of the effort has not been included in the cost. NASA only approved its new cost estimate through the initial operational readiness review, currently planned for September 2019. A project official said NASA headquarters asked the project to determine if there are ways to reduce the cost between the operational readiness review and the final acceptance review. NASA plans to conduct an independent review of the project in mid-2018 to inform a decision on whether to continue the project past the operational readiness review. If NASA decides to continue the project past this review, additional cost growth is expected for SGSS when NASA revisits project costs through future budget cycles.

NASA Generally Maintained Technology Maturity of Its Projects but Performance in Design Stability Measures Is Mixed

NASA has generally maintained the technology maturity and number of critical technologies of its major projects, but had mixed performance in design stability measures. Most of NASA's major projects in development matured their technologies to the level recommended by best practices by their preliminary design review—continuing a trend since 2015. Our best practices work has shown that reaching this level of maturity can minimize risks for projects entering development, which lowers the risk of subsequent cost growth and schedule delays. Of the three projects that held preliminary design review in

2017, one matured its technologies to the level recommended by GAO best practices. The average number of critical technologies decreased in 2018 because new projects in the portfolio reported one or no critical technologies. NASA has maintained the level of design stability at critical design review, but projects continued to experienced late design changes. Of the three projects that had held critical design review in 2017, one met the best practice of releasing 90 percent of engineering drawings by this review. In addition, the average percentage of released drawings for the portfolio—meaning the average of the percentage of drawings released for each project in the portfolio for each reporting period—has remained generally consistent with the past 5 years. We have previously found that releasing at least 90 percent of engineering drawings lowers the risk of projects experiencing design changes and manufacturing problems that can lead to cost and schedule growth.

NASA Generally Maintained the Technology Maturity of Its Projects

Most of NASA's major projects in development—12 of 17 projects—met the best practice of maturing all technologies to a technology readiness level 6 by their preliminary design review, which is a generally consistent trend since 2015 (see Figure 5).[24] Our best practices work has shown that reaching a technology readiness level 6—which includes demonstrating a representative prototype of the technology in a relevant environment that simulates the harsh conditions of space—by preliminary design review can minimize risks for the systems entering product development.[25] Projects falling short of this standard may experience subsequent technical problems, which can result in cost growth and schedule delays. Appendix IV provides a description of technology readiness levels, which are the metrics used to assess technology maturity.

This year, we removed one project—Origins-Spectral Interpretation-Resource Identification-Security-Regolith Explorer (OSIRIS-REx)—from our analysis of technology maturity because it launched in September 2016. We added three projects—Landsat 9, LCRD, and Restore-L— which held their preliminary design review before February 2018. Of the three projects that were added, only one—Landsat 9—matured its

[24] NASA distinguishes critical, or new, technologies from heritage technologies. Our product development best practices do not make this distinction. We describe critical technologies as those that are required for the project to successfully meet customer requirements, which can include both existing or heritage technology or new technology. Therefore, to assess overall technology maturity, we analyzed the maturity of heritage and critical technologies that NASA reported for projects in our data collection instrument. In other analyses, which focus on the number of new technologies being used by programs, we maintain NASA's distinction between critical and heritage technologies.

[25] Appendix V contains information about GAO's product development best practices and the project attributes and knowledge-based metrics that we assess projects against at each stage of a system's development.

technologies to a technology readiness level 6. The other two projects— Restore-L and LCRD—did not meet our technology maturity best practice.

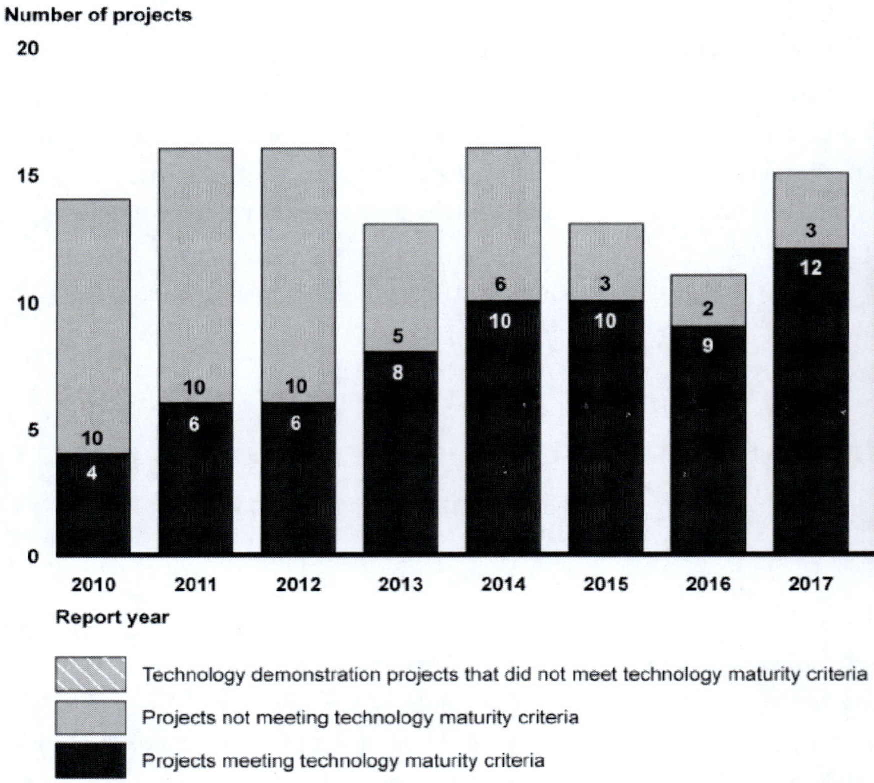

Source: GAO analysis of National Aeronautics and Space Administration data. | GAO-18-280SP.

Note: Includes projects that completed preliminary design review and identified critical or heritage technologies. For example, for 2018, 17 of 26 NASA major projects had held this review and identified critical or heritage technologies.

Figure 5. Number of NASA's Major Projects Attaining Technology Maturity by Preliminary Design Review from 2010 to 2018.

The LCRD and Restore-L projects are technology demonstrations managed by the Goddard Space Flight Center, and the center's policy does not require technology demonstrations to mature all of their technologies to technology readiness level 6 by preliminary design review.[26] NASA officials explained that this is because the purpose of some technology demonstration projects is to mature new technologies to technology readiness level 6 or higher by the end of the demonstration, making it not feasible for these projects to achieve this level by the preliminary design review. However, we included

[26] NASA's technology demonstration missions program, which began in 2010, aims to mature new technologies from a technology readiness level 5 to technology readiness level 7 or greater. After the technologies are matured, they are to be transferred or infused into other NASA, partner, or commercial projects.

LCRD and Restore-L in our analysis because both planned to mature their technologies prior to launching. Therefore, the same risks of subsequent technical problems that can result in cost growth and schedule delays identified in our best practices work applies to these projects. The LCRD project matured its technologies to technology readiness level 5 by its preliminary design review and to a technology readiness level 6 by its critical design review. The Restore-L project matured all but one of its technologies. The project changed its vision navigation system shortly before its preliminary design review because its prior vision navigation system did not meet requirements and the contractor was unable to resolve the issue. Restore-L project officials stated that they have a plan in place to mature the new system to a technology readiness level 6 by summer 2018, prior to the project's planned critical design review in January 2019.

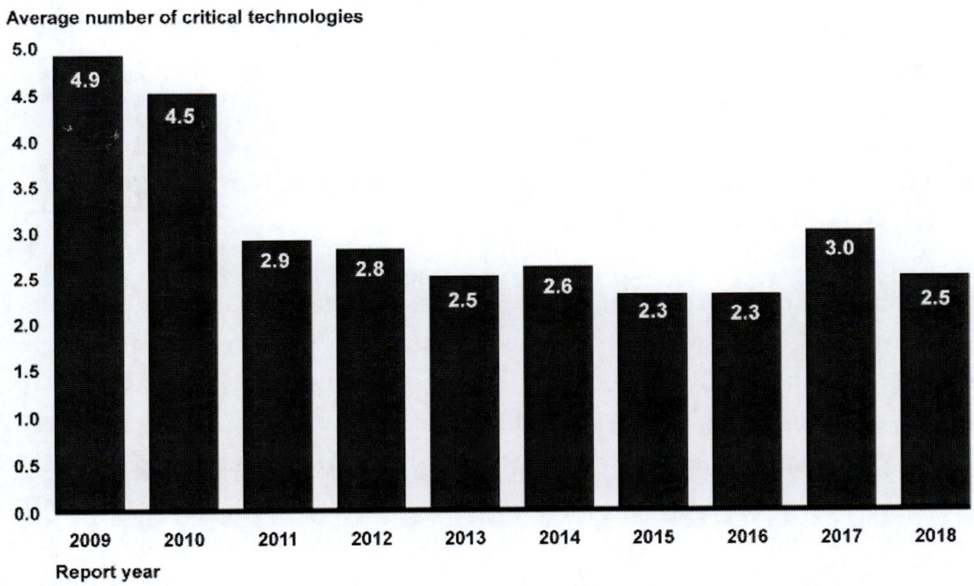

Source: GAO analysis of National Aeronautics and Space Administration data. | GAO-18-280SP.

Figure 6. Average Number of Critical Technologies Reported by NASA's Major Projects in Development from 2009 to 2018.

NASA's major projects continue to report low numbers of critical technologies.[27] The average number of critical technologies for the projects in development declined to 2.5 compared to last year's average of 3 critical technologies; however, the average is consistent with the last several years (see Figure 6). The decrease in critical technologies was driven by the addition of three projects—Landsat 9, LCRD, and RestoreL—that reported one or no critical technologies and the exclusion of OSIRIS-REx, which had 3 critical technologies.

[27] Projects self-report the number of their critical technologies.

NASA's most recent Discovery project selections—Lucy and Psyche— continue its trend of low reported critical technologies, but are not yet included in our analysis because they have not held preliminary design review. These projects reported no critical technologies and plan to use a design that is based heavily on heritage technologies with modifications. Both projects reported that their heritage technologies are mature, but plan to modify some technologies. We have previously found that mature technologies must be demonstrated in a relevant environment and should be very close to form, fit, and function. Therefore, these two projects will likely have developmental work despite using high-heritage, mature technologies in their designs.

GAO's best practices criteria do not focus on the number of new technologies, but rather their maturity, when considering their effect on cost and schedule risk. Therefore, the issue is not whether to push innovation through technology development, but rather the steps projects take to increase the likelihood of mission success by maturing these technologies to a high level prior to entering the implementation phase.

NASA Has Maintained the Level of Design Stability at Key Design Review, but Late Design Changes Still Occurred

NASA has maintained the number of projects with stable designs at critical design review. The critical design review is the time in the project's life cycle when the integrity of the project design and its ability to meet mission requirements is assessed. Our work on product development best practices shows that at least 90 percent of engineering drawings should be releasable by this review to lower the risk of subsequent cost and schedule growth. Engineering drawings are considered to be a good measure of the demonstrated stability of a product's design because the drawings represent the language used by engineers to communicate to the manufacturers the details of a new product design—what it looks like, how its components interface, how it functions, how to build it, and what critical materials and processes are required to fabricate and test it. Once the design of a product is finalized, the drawing is "releasable."

Of the 12 projects that have held a critical design review, four released at least 90 percent of their current projected engineering drawings (see Figure 7).[28] In addition, the average percentage of released drawings for the portfolio has remained generally consistent with the past 5 years.

[28] We used the number of drawings released at the critical design review as a fraction of the total number of drawings currently projected. Therefore, the denominator in the calculation may have been larger than what was projected at the critical design review. We believe this calculation more accurately reflects the design stability of the project.

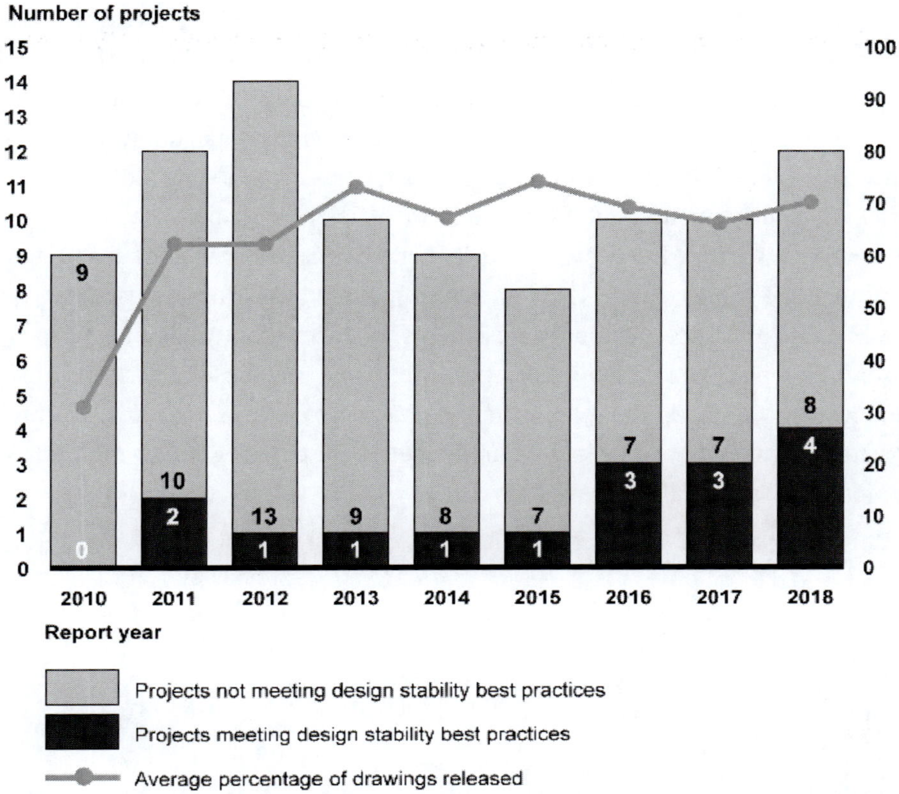

Source: GAO analysis of National Aeronautics and Space Administration data. | GAO-18-280SP.

Note: Includes projects that completed critical design review and had engineering drawings. For example, for 2018, 12 of 26 NASA major projects had held this review and had engineering drawings.

Figure 7. Number of NASA Major Projects That Released at least 90 Percent of Drawings by Critical Design Review and Average Percentage of Released Drawings from 2010 to 2018.

This year, in analyzing the number of projects that met the best practice, we added 3 projects and removed 1 project. One of the 3 projects that were added to the analysis this year—RBI—met the best practice. The other two projects, LCRD and Mars 2020, did not meet the best practice. The LCRD project released over 90 percent of its design drawings in December 2016 at its critical design review, but reported design drawing growth in January 2018. As a result of the drawing growth, the metric of drawings released at this review decreased to 78 percent, which is below the best practice. The Mars 2020 project held its critical design review before it had a stable design, at which point it had released about 72 percent of its expected design drawings. Project officials stated that they held the design review earlier than normal to avoid delaying the development of heritage technologies, which make up a large percentage of the Mars 2020 rover. As a result of holding the review early, new and highly complex developments were not stable at the time of the design review.

Our analysis of design drawings dropped one project—OSIRIS-REx—in 2018 and removed the EGS program from prior years' analyses. We excluded EGS from prior years because program officials told us that the prior data submissions were not inclusive of all subsystems and they did not have a way to replicate previous data submissions to include all of the subsystems. Further, EGS program officials stated they do not use design drawings as a management tool. EGS program officials said that the program used a different approach for tracking design maturity because the program contains several major construction and facilities efforts. This approach included tracking requirements stability, internal and cross-program technical assessments, interface control document maturation, and closure of open work. However, because some of these efforts required new development work, tracking the release of and changes to engineering design drawings at the program level would have been a valuable metric for the program to monitor. We will follow up on this issue in a future review.

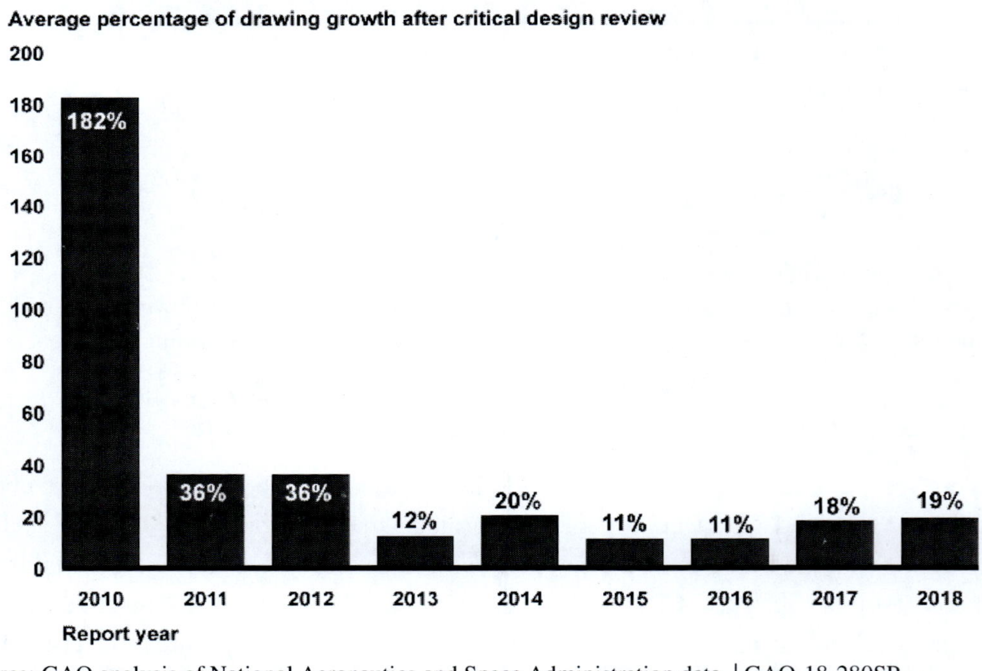

Source: GAO analysis of National Aeronautics and Space Administration data. | GAO-18-280SP.

Note: Drawing growth in 2010 was primarily attributed to the Solar Dynamics Observatory (SDO) because it did not have a stable design at its critical design review and drawings for SDO's instruments were not included in this review. The project launched in 2010 and exited the portfolio.

Figure 8. Average Percentage of Engineering Drawing Growth after Critical Design Review for NASA Major Projects from 2010 to 2018.

Despite an increase in the number of projects meeting GAO's best practice, some of the 8 projects that did not meet the best practice released a low percentage of their drawings by the critical design review. Five of the 8 projects that did not meet the best practice released fewer than 60 percent of their projected engineering drawings. For example, in 2016 we found that the Parker Solar Probe project held its critical design review with only 34 percent of its design drawings released, significantly lower than the best practice of 90 percent.[29] The project demonstrated it had a stable design at its system integration review in May 2016—over one year after its critical design review, the point at which product development best practices recommend having a stable design.

In addition to maintaining the level of design stability at critical design review, NASA major projects also experienced design changes at a similar level as last year. The average percentage of drawing growth for the entire portfolio slightly increased from 18 percent to 19 percent (see Figure 8). Of the 12 projects included in this year's analysis, 11 experienced drawing growth after their critical design review, ranging from 2 to 52 percent. If a project experiences a large amount of engineering drawing growth after this review, it may be an indicator of instability in the project design late in the development cycle. Design changes at this point can be costly to the project in terms of time and funding because hardware may need to be reengineered or reworked as a result. This analysis dropped OSIRIS-REx in 2018 and the EGS program in prior years for reasons discussed above.

NASA projects have continued to perform well against other design stability measures, such as maintaining center-specified mass and power margins, and completing a validation and verification plan.[30] Most projects for which mass and power requirements were applicable met these requirements at their critical design reviews and have continued to maintain the required levels of margin.[31] Further, all projects that were required to complete a validation and verification plan by their critical design review met the requirement. NASA requires that projects complete a validation and verification plan by this review to ensure that projects have a plan in place to track the completion of verification and validation events and activities.

Too Early to Determine Outcome of NASA's Efforts to Address Workforce Challenges

Several projects covered in our review were experiencing workforce challenges, such as not having enough staff or staff with the right skills. While it is difficult to tie these

[29] GAO-16-309SP.

[30] Validation is defined as the continuous process of ensuring that requirements are well-formed (clear and unambiguous), complete (agrees with customer and stakeholder needs and expectations), consistent (conflict free), and individually verifiable and traceable to a higher level requirement or goal. Verification is defined as proof of compliance with requirements and specifications.

[31] Mass is a measurement of how much matter is in an object. It is related to an object's weight, which is mathematically equal to mass multiplied by acceleration due to gravity. Margin is the spare amount of mass or power allowed or given for contingencies or special situations. Some centers provide guidance on the percentage of mass margin required at various points in project development, with required margins ranging from 30 to 0 percent, depending on where a project is in the development cycle.

workforce challenges to increases in cost, there have been instances where projects have reported schedule delays or late design changes as a result of workforce shortages. Our work has generally found that the shortage of trained acquisition personnel hinders agencies from managing and overseeing acquisition programs and contracts that have become more expensive and increasingly complex.[32] Moreover, NASA's own assessments indicate that there are broader workforce-related challenges that can have a negative impact on programs over the long run, if not addressed in a strategic manner. NASA is taking steps to address its findings, but it is too early to determine the outcomes. One is focused on developing a strategic workforce plan, including determining what mix of skills and capabilities are needed in the future, and the other is focused on leveraging skills across NASA's centers. We will be tracking NASA's progress in this area in future reviews as capacity—having the right people and resources to resolve risk—is one of two criteria that NASA has not yet met for removal from our high-risk list.[33]

NASA Has Experienced Workforce Challenges on Major Projects

During the course of our review, we found that several major projects identified workforce challenges, including not having enough staff or staff with the right skills. To accomplish its scientific and space exploration missions, NASA relies on having a skilled civil servant and contractor workforce. While it is difficult to tie these workforce challenges to increases in cost, there have been instances where projects have reported schedule delays or late design changes as a result of workforce shortages. Further, our work has generally found that the shortage of trained acquisition personnel hinders agencies from managing and overseeing acquisition programs and contracts that have become more expensive and increasingly complex.[34] Examples of workforce challenges identified by projects include:

- The Mars 2020 project, which is managed by the Jet Propulsion Laboratory, experienced workforce shortages in several key areas such as flight software and for the development of a sampling and caching system that will collect and cache Martian soil and rock samples. Project officials said not having staff with the right level of experience within its sampling and caching system development contributed to the project being behind in releasing its engineering drawings. As of June 2017, project officials said the Jet Propulsion Laboratory has been able to

---

[32] GAO, *Acquisition Workforce: Federal Agencies Obtain Training to Meet Requirements, but Have Limited Insight into Costs and Benefits of Training Investment,* GAO-13-231 (Washington, D.C.: Mar. 28, 2013).
[33] GAO-17-317.
[34] GAO-13-231.

address these workforce challenges largely because laboratory management prioritized staffing for the project over other projects.

- The Europa Clipper project, which is managed by the Jet Propulsion Laboratory, experienced workforce shortages in several key areas, including avionics, since October 2016. The project has revised staffing plans several times, but continues to face workforce shortages even with NASA headquarters involvement. Project officials said these shortages have delayed development in some areas and have led the project to outsource some work to contractors. As of January 2018, project officials reported that they have seen improvements in addressing workforce shortages.
- The SGSS project, managed by the Goddard Space Flight Center, faced staffing shortfalls in key areas, such as systems engineering and business management, which contributed to late design changes. For example, a project official said the systems engineering was not done correctly before, which led to the software not being coded to the right requirements. As of January 2018, NASA officials stated that these workforce issues had largely been resolved because some positions were filled by reassigning staff to these areas and others are no longer needed.
- The EGS program, which is managed by Kennedy Space Center, is working with its contractors to increase its software development staff retention rate, which has been a challenge. Program officials said the current retention rate is about 70 percent, but the program would like that to be higher in order to avoid additional delays in its software development efforts due to losing staff. The officials attributed the contractor workforce challenges to competition in the space industry near the center in Florida.

Further, in January 2018, the NASA Office of Inspector General found that the SWOT project, which is managed by the Jet Propulsion Laboratory, experienced workforce shortages in key fields to design and develop its instruments.[35] The report stated that the staffing problems were driven by conflicting priorities among several other Jet Propulsion Laboratory projects, including Mars 2020, that also needed those skills. The Office of Inspector General recommended that the Jet Propulsion Laboratory director evaluate current and future critical technical staffing requirements and make adjustments as necessary to ensure that the laboratory has the technical support needed for their missions. The director plans to address the recommendation through its existing processes and forums for assessing and addressing staffing resources at the laboratory.

---

[35] NASA Office of Inspector General, *NASA's Surface Water and Ocean Topography Mission*, IG-18-011 (Washington, D.C.: Jan. 17, 2018).

NASA's Assessments Indicate Broader Workforce Challenges

NASA's own assessments indicate that there are broader workforce challenges in disciplines that are necessary to effectively manage its major projects. Since 2012, NASA has been conducting capability assessments of its technical disciplines and business services to assist in establishing a more efficient operating model that maintains critical capabilities and meets current and future mission needs. Through these efforts, NASA identified capacity gaps in the following areas:

- **Guidance, Navigation, and Control (GN&C) disciplines.** An assessment of technical capabilities conducted in 2016 identified GN&C research and technology development as a discipline area that was inadequately positioned to meet mission requirements. In a follow-up assessment in 2017, NASA determined there have been no significant changes to the GN&C workforce since 2015. The assessment found, though, that the risk is stabilizing due, in part, to adding more early career hires. These assessments determined the GN&C workforce needs to be reshaped over the next decade to build future workforce skills.
- **Scheduling, earned value management, and cost estimating.** A budget business services assessment conducted in 2016 recommended that the agency work with centers to regionalize access to these specialty capabilities and to leverage access to these skills across centers. Officials from the Office of the Chief Financial Officer said they have made some progress in filling gaps in cost estimating and earned value management capabilities, but scheduling expertise remains a major challenge. As a result, officials said they have a scheduling effort underway to build the capability in-house by improving training and tools.
- **Cost and pricing analyst workforce.** A procurement business services assessment conducted in 2015 identified capability gaps in the cost and pricing analyst workforce. As part of assessing options to strengthen expertise associated with source evaluation boards for contract awards, NASA found that the agency had an inadequate supply of cost and pricing analysts across centers. In response to this finding, NASA plans to form a centralized team of cost and pricing professionals to support acquisition efforts across centers. In addition, NASA established a pricing community of practice to develop and share best practices in this area, developed training courses, and, in some instances, centers are hiring cost and pricing analysts as a part of efforts to rebuild this skill set across the agency.

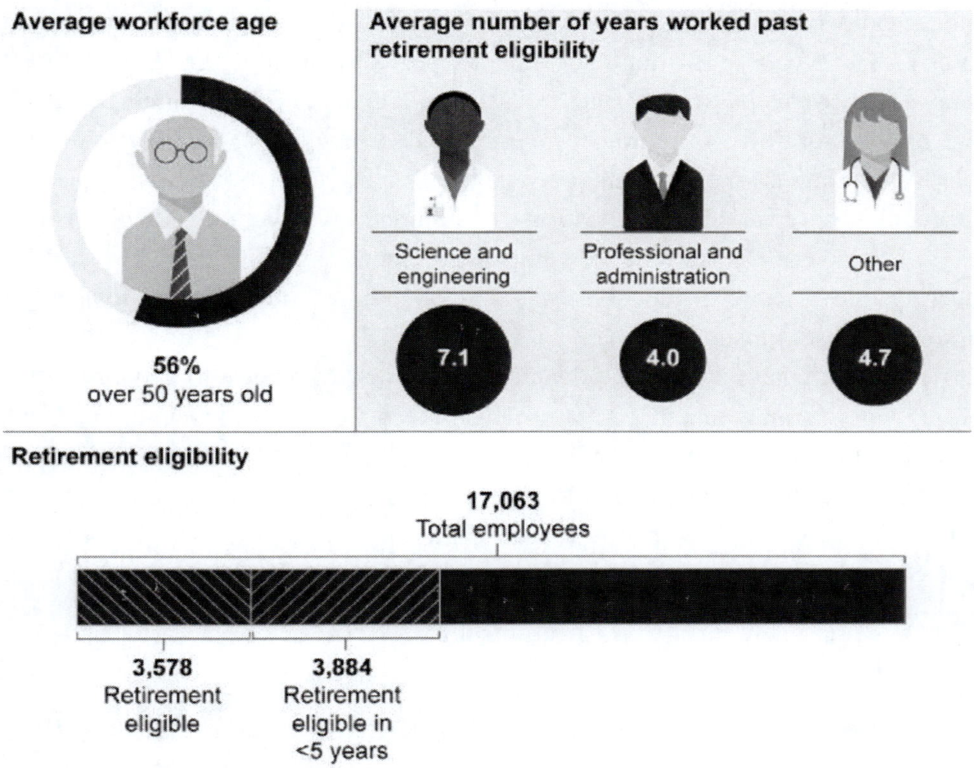

Source: GAO analysis of National Aeronautics and Space Administration data. | GAO-18-280SP.
Note: The other category includes clerical, technicians, medical and other miscellaneous occupations.

Figure 9. NASA's Workforce Average Age, Retirement Eligibility, and Average Number of Years Worked Past Retirement Eligibility as of January 2018.

We have also found that other agencies have experienced challenges related to their earned value management, cost estimating, and cost and pricing analysis workforces.[36] For example, in June 2008, we found that the Coast Guard faced challenges in recruiting and retaining a sufficient number of government employees in acquisition positions such as contract specialists and cost estimators, among other areas.[37] In addition, in December 2015, we found that the Department of Defense did not meet its workforce growth goals in 3 priority career growth areas— contracting, business, which includes cost estimating and analysis, and engineering.[38]

Another trend, the aging of NASA's workforce, has both negative and positive effects. About 56 percent of NASA's workforce is 50 years old and over, an increase of about 7

---

[36] For example, GAO, *Coast Guard: Change in Course Improves Deepwater Management and Oversight, but Outcome Still Uncertain,* GAO-08-745 (Washington, D.C.: June 24, 2008); *Defense Contract Management Agency: Amid Ongoing Efforts to Rebuild Capacity, Several Factors Present Challenges in Meeting Its Missions,* GAO-12-83 (Washington, D.C.: Nov. 3, 2011); and *Defense Acquisition Workforce: Actions Needed to Guide Planning Efforts and Improve Workforce Capability*, GAO-16-80 (Washington, D.C: Dec. 14, 2015).

[37] GAO-08-745.

[38] GAO-16-80.

percentage points over the past 5 years. Officials said that NASA's workforce is aging because NASA has a low attrition rate—about 4 percent annually—and high numbers of staff stay several years past retirement. We also found that, as of the beginning of 2018, 21 percent of the workforce is retirement eligible, about another 23 percent will become eligible in less than 5 years, and the average number of years staff that stay past initial retirement eligibility varied by occupation (see Figure 9). On average, individuals remain at NASA between 4-7 years past their initial retirement eligibility date, but staff in the engineering and science occupations stay on longer than other occupations, such as professional and administrative.

Officials said there are both advantages and disadvantages to having an aging workforce. For example, human capital officials noted that having an aging workforce is good for maintaining institutional knowledge due to experienced staff staying longer, but that having a low attrition rate makes it more difficult for the agency to make changes to its workforce skill mix as needed. Officials within the Office of the Chief Engineer and Mission Support Directorate said that they were looking at ways to be more strategic in hiring or using existing capabilities to meet their skills needs.

Outcome of NASA's Strategic Workforce and Cross-Center Planning Efforts Not Yet Known

In our review of selected agency-wide assessments and through discussions with knowledgeable NASA officials, we found that NASA has two efforts underway that may help to address some workforce challenges: strategic workforce planning and cross-center collaboration. We will be tracking the progress of these efforts in future reviews, as it is too early to determine the outcomes.

Strategic Workforce Planning

In 2016, NASA completed a human capital assessment that identified a lack of agency-wide strategic workforce planning as a high-risk area under the agency's current operating model. It stated that the lack of agency-wide strategic workforce planning was at the root of many of the other findings in the assessment, such as the need for better coordination and collaboration across the centers, created uncertainty and was driving centers to think locally instead of agency-wide. To address this concern, the Office of Human Capital Management has an effort underway to begin to develop NASA's first agency-level strategic workforce plan in 2018. NASA defines strategic workforce planning as the discipline of determining the size and composition of a future workforce that is able to perform the organization's most important functions, maintain capabilities, and fulfill key business goals while maintaining agility in a dynamic environment. Human capital officials said they plan to update NASA's policy on strategic workforce planning after they have established their framework for strategic workforce planning.[39] As part of the new process, human capital officials said their office will provide guidance to the centers to develop center master plans, which are to have a 10- to15-year outlook on capabilities. Human capital officials said they will ask the centers to develop their first master plans by October 2018, and the Office of Human Capital Management is to use them to create NASA's first agency-level strategic workforce plan.

Cross-Center Collaboration

The 2016 human capital assessment also identified the need to increase cross-center collaboration to improve workforce utilization. NASA plans to develop a process to better understand where skills are needed across the centers and then use employees with those skills at any center to meet those needs. According to human capital officials, the Office of Human Capital Management was not the right organization to address this concern because they felt that it would require leadership at a higher level than their office to implement it. They noted that an effort such as this would require the agency to overcome the cultural barrier of NASA being a decentralized agency, with most of its work completed at the center level. This barrier has proven difficult for the agency in the past and has resulted in NASA having different approaches to implementing policies and guidance at the centers. For example, in April 2014, we found that when NASA policy gave centers wide latitude in implementing export control procedures, implementation across centers was inconsistent.[40]

NASA has taken other steps to improve cross-center collaboration by sharing mission support capabilities across the centers. For example, NASA plans to coordinate the center-level human capital and financial staff and activities through the headquarters offices, rather than at each center. NASA plans to complete this transition by the end of September 2019.[41] Other business services areas, such as procurement, also plan to move to this operational model, but have not yet started the transition. Officials within the Office of the Chief Financial Officer said the purpose of this change is for the mission support areas to become more effective and efficient, and deliver a consistent set of services across the agency.

As NASA develops its first strategic workforce plan and considers taking additional steps towards a new operating model with more cross-center collaboration, it will be vital to have senior leadership's commitment to these efforts. Capacity—having the right people and resources to resolve risk—is one of two criteria that NASA has not yet met for removal from our high-risk list; the other being demonstrating progress. We first designated acquisition management at NASA as a high-risk area in 1990 in view of NASA's history

---

[39] NASA Policy Directive 3010.1A, *Strategic Workforce Planning* (Jan. 15, 2009).

[40] GAO, *Export Controls: NASA Management Action and Improved Oversight Needed to Reduce the Risk of Unauthorized Access to Its Technologies,* GAO-14-315 (Washington, D.C.: Apr. 15, 2014).

[41] The Office of the Chief Financial Officer capabilities that are not realigned under the new model before October 2019 will be handled as part of a separate effort.

of persistent cost growth and schedule slippage in the majority of its major systems. Capacity is central to managing the complex and difficult work that NASA takes on. We have found that an effective organization includes a senior leadership team committed to developing more effective ways of doing business, accomplishing results, and investing in human capital.[42]

PROJECT ASSESSMENTS

The individual assessments of the 26 projects we reviewed provide a two-page profile of each project. Each assessment includes a description of the project's objectives, information about the NASA centers and international partners involved in the project, the project's cost and schedule performance, a timeline identifying key project dates, budget information, and a brief narrative describing the current status of the project.[43] The budget information is based on NASA's fiscal year 2019 budget request. On the first page, the project profile presents the standard information listed above. On the second page of the assessment, we provide an analysis of the project challenges, and outline the extent to which each project faces cost, schedule, or performance risks because of these challenges, if applicable. NASA project offices were provided an opportunity to review drafts of the assessments prior to their inclusion in the final product, and the projects provided both technical corrections and more general comments. We integrated the technical corrections as appropriate and summarized the general comments at the end of each project assessment.

See Figure 10 for an illustration of a sample assessment layout.

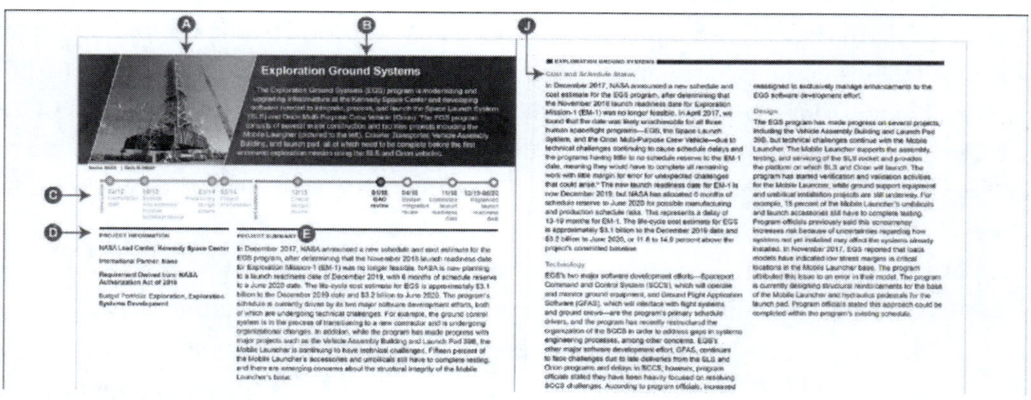

Figure 10. (Continued).

[42] GAO, *Exposure Draft: A Model of Strategic Human Capital Management,* GAO-02-373SP (Washington, D.C.: Mar. 15, 2002).

[43] The manifested launch date is the launch date which the project is working toward, and when a launch vehicle is available to launch the project. This date is only a goal launch date for the project, not a commitment that they will launch on this date. The committed launch readiness date is determined through a launch readiness review that verifies that the launch system and spacecraft/payloads are ready for launch.

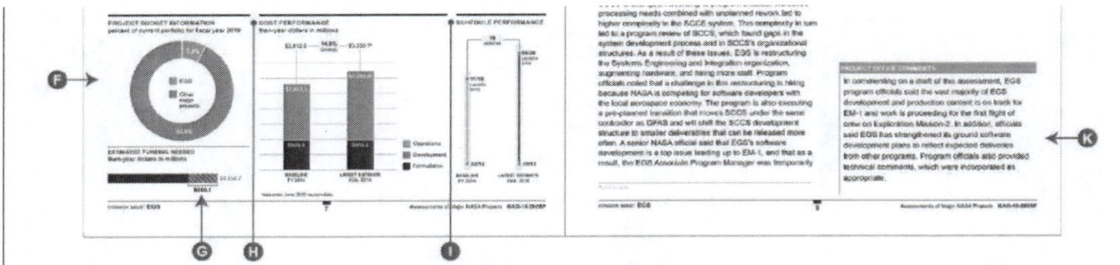

- **A** Illustration of the spacecraft, instrument, aircraft, launch vehicle, or ground system.
- **B** General description of the mission's objectives.
- **C** Timeline identifying key dates for the project including when the project began formulation, held major design reviews, began implementation, and launched or fielded an operating capability.
- **D** **Project Information:** Information on the responsible NASA center, international partners, launch plans, mission duration, the source of mission's requirements, and budget portfolio.
- **E** **Project Summary:** Brief narrative describing the current status of the project.
- **F** **Project Budget Information:** The percentage of NASA's fiscal year 2019 budget for all major projects in the current portfolio that the project represents.
- **G** **Project Funding:** For projects in formulation, the total funding appropriated to the project since formulation start. For projects in implementation, the funding needed to be allocated for project completion as of the latest life-cycle cost estimate.
- **H** **Cost:** For projects in formulation, the preliminary cost estimate. For projects in implementation, the approved cost baseline and latest estimate.
- **I** **Schedule:** For projects in formulation, the preliminary launch readiness target date or range of dates. For projects in implementation, the approved schedule baseline and latest estimate.
- **J** The second page of the assessment is an analysis of the project challenges and the extent to which each project faces cost, schedule, or performance risks because of these challenges.
- **K** **Project Office Comments:** General comments provided by the cognizant project office.

Source: GAO analysis. | GAO-18-280SP.

Figure 10. Illustration of a Sample Project Assessment.

Commercial Crew Program

The Commercial Crew Program facilitates and oversees the development of safe, reliable, and cost-effective crew transportation systems by commercial companies to carry NASA astronauts to and from the International Space Station (ISS). The program is a multi-phase effort that started in 2010. During the current phase, the program is working with two contractors—Boeing and SpaceX—that will design, develop, test, and operate the crew transportation systems. Once NASA determines the systems meet its standards for human spaceflight—a process called certification—the companies will fly up to six crewed missions to ISS.

Source: NASA | GAO-18-280SP

FORMULATION

- 09/14 Transportation Capabilities phase contract awards
- **01/18 GAO review**
- 08/18 SpaceX uncrewed test flight
- 08/18 Boeing uncrewed test flight
- 11/18 Boeing crewed test flight
- 12/18 SpaceX crewed test flight
- 01/19 Boeing final certification
- 02/19 SpaceX final certification
- 2019 – 2024 Post-certification missions

PROJECT INFORMATION

NASA Lead Center: **Kennedy Space Center**

Commercial Partners: **Boeing, SpaceX, Blue Origin,[a] Sierra Nevada Corporation[a]**

Launch Location: **Boeing-Cape Canaveral Air Force Station, FL; SpaceX-Kennedy Space Center, FL**

Launch Vehicle: **Boeing-Atlas V; SpaceX-Falcon 9**

Requirement Derived from: **NASA Strategic Plan**

Budget Portfolio: **Space Operations, Space Transportation**

[a]Blue Origin and Sierra Nevada Corporation do not have contracts for the current phase and therefore were not included in this assessment.

PROJECT SUMMARY

Both of the Commercial Crew Program's contractors have made progress developing their crew transportation systems, but continue to have aggressive development schedules. All test flights have slipped to 2018 and the final certification reviews have slipped to early 2019. Both contractors continue to work through design issues for their crew transportation systems. SpaceX is finalizing the design for its launch vehicle, Falcon 9, which will be needed to support crewed flights. Boeing is analyzing the design of the forward heat shield to determine if its expected performance may damage the spacecraft during re-entry. The Commercial Crew Program is the first NASA program that the agency will evaluate against a loss of crew requirement, a key safety metric. Program officials said that if the contractors cannot meet the loss of crew requirement specified in the contracts, NASA could still certify their systems by employing operational mitigations.

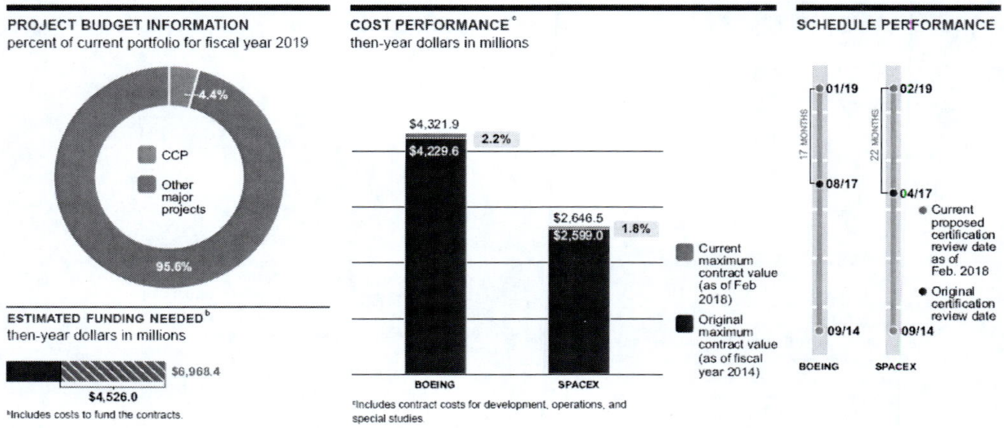

Cost and Schedule Status

Both of the Commercial Crew Program's contractors have made progress developing their crew transportation systems, but delays persist as the contractors have had difficulty executing aggressive schedules. The contractors were originally required to provide NASA all the evidence it needed to certify that their systems met its requirements by 2017. In January 2018, we found the contractors' test flights have slipped to 2018 and the final certification reviews have slipped to early 2019.[44] This represents a delay of 17 months for Boeing and 22 months for SpaceX from initial schedules. The Commercial Crew Program is tracking risks that both contractors could experience additional schedule delays and its schedule risk analysis indicates that certification is likely to slip until late 2019 for SpaceX and early 2020 for Boeing.

Design

SpaceX must close several of the program's top risks related to its upgraded launch vehicle design, the Falcon 9 Block 5, before it can be certified for human spaceflight. This includes SpaceX's redesign of the composite overwrap pressure vessel. SpaceX officials stated the new design aims to eliminate risks identified in the older design, which was involved in an anomaly that caused a mishap in September 2016. Separately, SpaceX officials told us that the Block 5 design also includes design changes to address cracks in the turbine of its engine identified during development testing. NASA program officials

[44] GAO-18-317T.

told us that they had informed SpaceX that the cracks were an unacceptable risk for human spaceflight. SpaceX officials told us that they have made design changes, captured in this Block 5 upgrade, that did not result in any cracking during initial life testing. However, this risk will not be closed until SpaceX successfully completes qualification testing in accordance with NASA's standards without any cracks. SpaceX officials stated they expect this testing to be completed in first quarter calendar year 2018.

Boeing is also mitigating several risks in order to certify its system including challenges related to the performance of its abort system and a component of its parachute system. Boeing is addressing a risk that its abort system may not meet the program's requirement to have sufficient control of the vehicle through an abort. In some abort scenarios, Boeing has found that the spacecraft may tumble, which could pose a threat to the crew's safety. To validate the effectiveness of its abort system, Boeing has conducted extensive wind tunnel testing and plans to complete a pad abort test in April 2018. Boeing is also addressing a risk that during re-entry to the Earth's atmosphere, a portion of the spacecraft's forward heat shield, which protects the parachutes during re-entry, may reconnect and damage the parachute system. NASA's independent analysis indicates that this may occur if both parachutes that pull the forward heat shield away from the spacecraft deploy as expected. Boeing's analysis indicates the risk exists only if one of two parachutes does not deploy as expected. If the program determines this risk is unacceptable, Boeing would need to redesign the parachute system, which the program estimates could result in at least a 6-month delay.

Other Issues to Be Monitored

The ability to meet one of the crew safety requirements, which stems from agency policy, is one of the program's top risks. NASA established the "loss of crew" requirement as one way to measure the safety of a crew transportation system, with the intent to minimize the probability of death or permanent disability to one or more crew members. Under the current contracts, the loss of crew requirement is 1 in 270, meaning that the contractors' systems must carry no more than a 1 in 270 probability of incurring loss of crew. Program officials told us that Commercial Crew is the first NASA program that the agency will evaluate against a loss of crew requirement. They said that if the contractors cannot meet the loss of crew requirement, there are several actions the Commercial Crew Program could take to help meet it. Program officials stated, however, these actions may not be enough to completely close the gap. The program has reported it is exploring options, such as on-orbit inspections of the spacecraft. Additionally, program officials told us that one of their greatest upcoming challenges will be to complete two oversight activities— conducting phased safety reviews and verifying that contractors meet requirements—concurrently. For example, the program originally planned to complete phase two of the safety review process in early 2016, but as of October 2017, neither contractor had completed this phase—the program had approved 90 percent of Boeing's phase two reports and 70 percent of SpaceX's.

Project Office Comments

In commenting on a draft of this assessment, program officials stated that NASA's Commercial Crew Program and private industry partners, Boeing and SpaceX, continue to develop the systems that will return human spaceflight to the United States. Both commercial partners are undertaking considerable amounts of testing in 2018 to prove space system designs and the ability to meet NASA's safety and mission requirements for crew flights to the ISS.

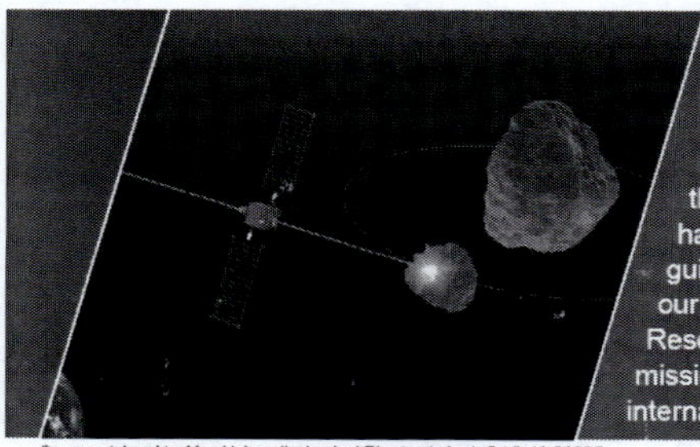

Double Asteroid Redirection Test

The DART project plans to travel to the near-Earth asteroid Didymos, a binary system, and impact the smaller of the two bodies. NASA will assess the deflection result of the impact for potential future use on other potentially hazardous near-Earth objects. The project responds to near-Earth object guidance by the Office of Science and Technology Policy to better understand our impact mitigation posture, and to recommendations by the National Research Council Committee to conduct a test of a kinetic impactor. The DART mission is part of the Asteroid Impact and Deflection Assessment, which is an international collaboration with the European Space Agency.

Source: Johns Hopkins University/Applied Physics Lab. | GAO-18-280SP

FORMULATION
- 02/16 Formulation start
- 09/16 System requirements/ mission definition review
- **01/18 GAO review**
- 04/18 Preliminary design review
- 05/18 Project confirmation

IMPLEMENTATION
- 03/19 Critical design review
- 11/19 System integration review
- 12/20 Projected launch readiness date

PROJECT INFORMATION

NASA Lead Center: **Marshall Space Flight Center**

International Partner: **European Space Agency**

Launch Location: **TBD**

Launch Vehicle: **TBD**

Mission Duration: **22 months**

Requirement Derived from: **NASA Authorization Act of 2008 and implementing guidance**

Budget Portfolio: **Science, Planetary Science**

PROJECT SUMMARY

The DART project is working toward preliminary design review and project confirmation, but funding uncertainty and technology maturity of its propulsion system may change the project's schedule. This uncertainty could trigger a re-evaluation of life-cycle costs and overall schedule heading to preliminary design review and project confirmation. Further, DART's planned electric propulsion thruster is facing development delays, particularly in its propulsion power systems. The project is looking at possible schedule delays should thruster issues persist. DART was also intended to be one of two spacecraft to be sent to the Didymos binary asteroid system—along with a spacecraft produced and launched by the European Space Agency. The European spacecraft did not receive necessary funding for 2017 and may be canceled. The DART project is assessing the effect on collected science data of losing proximate observation and is in the early stages of considering various alternatives.

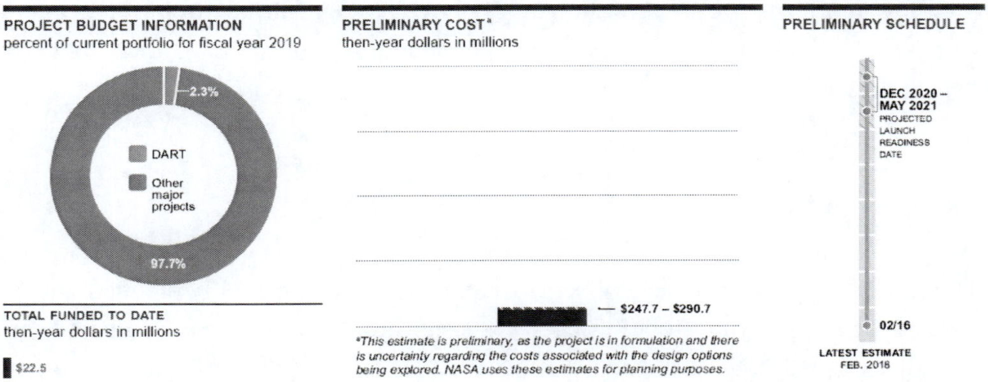

Cost and Schedule Status

The DART project entered the preliminary design and technology completion phase in June 2017 with the agency acknowledging that uncertain funding may affect the project's schedule. This could trigger a re-evaluation of life-cycle costs and overall schedule heading to preliminary design review and project confirmation—which the project has delayed by 2 months and 1 month, respectively. DART's planned launch readiness date range is December 2020 – May 2021, with impact of the asteroid occurring in 2022. Project officials told us that if they receive less funding than planned, there is another launch opportunity 2 years later and they are currently studying the effects on the mission if it is delayed to this date. Officials also noted a delayed launch window would likely increase total costs in exchange for lower annual costs.

Technology

DART's propulsion system includes technologies that remain immature and may delay the mission preliminary design review and potentially the project's launch date. The DART project is working with NASA's Evolutionary Xenon Thruster-Commercial (NEXT-C)—a project managed at Glenn Research Center under the Discovery Program—to provide in-space electric propulsion for the mission. The DART mission will be the first time this technology will fly operationally and, if successful, will serve as qualification for future deep space missions. However, first-time production of this system—DART's critical path and top risk—is proving challenging. The NEXT-C thruster's power processing unit, which provides necessary voltages to the thruster, is the technical risk keeping the thruster from maturing. The project has delayed the mission preliminary design review by 2 months in part to allow more time to mature the NEXT-C technology. In addition, the program is carrying a risk to its launch readiness date dependent upon the NEXT-C meeting delivery and design targets.

Developmental Partner

The DART project's science collection may be less than originally planned due to the European Space Agency not funding DART's sister project, the Asteroid Impact Monitor (AIM) project. DART was designed to work with AIM as part of the collaborative Asteroid Impact & Deflection Assessment. Per the mission plan, as the DART spacecraft impacts the smaller of the Didymos binary asteroids, the AIM spacecraft would collect impact data along with Earth-based observatories. However, the AIM project did not receive necessary funding at the European Space Agency's ministerial meeting in December 2016. Losing AIM will reduce the amount of science data collected on the impact and the asteroids themselves. While the project believes that it can still meet its objectives of characterizing the results of the DART impact on the Didymos system, the project is performing planning and analysis leading up to preliminary design review to determine how effective the Earth-based observatories will be at determining the results of the impact event. As a result of the potential loss of science data, NASA is considering possible alternatives that could provide imagery and data related to the impact.

Project Office Comments

DART project officials provided technical comments on a draft of this assessment, which were incorporated as appropriate.

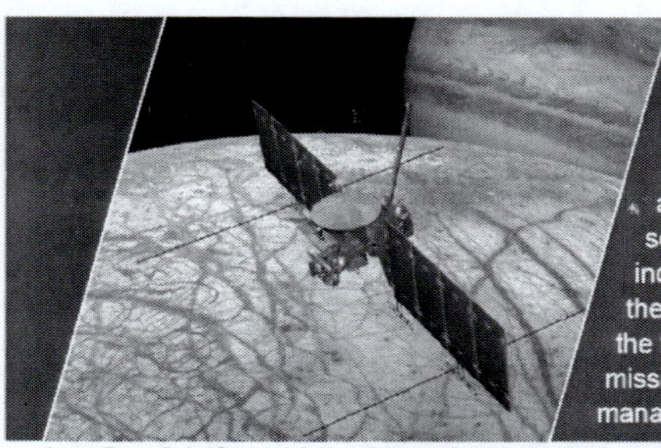

Europa Clipper

The Europa Clipper mission aims to investigate whether the Jupiter moon could harbor conditions suitable for life. The project plans to launch a spacecraft in the 2020s, place it in orbit around Jupiter, and conduct a series of investigatory flybys of Europa. The mission's planned objectives include characterizing Europa's ice shell and any subsurface water, analyzing the composition and chemistry of its surface and ionosphere, understanding the formation of its surface features, and surveying sites for a potential landed mission. We did not assess the proposed Europa lander mission, which NASA is managing as a separate project in pre-formulation.

Source: Europa Project Personnel, California Institute of Technology, Jet Propulsion Laboratory. | GAO-18-280SP

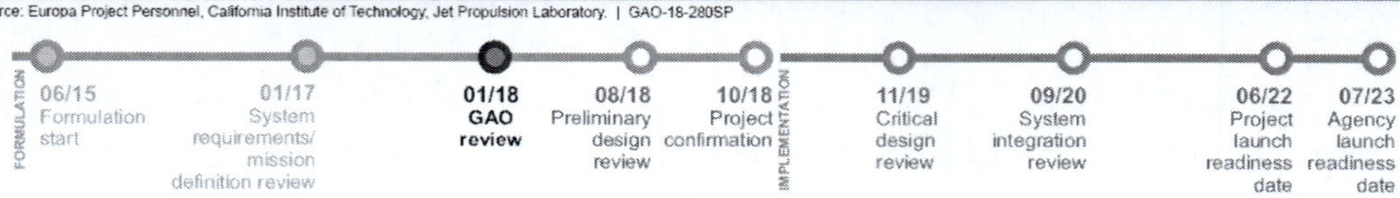

PROJECT INFORMATION

NASA Lead Center: **Jet Propulsion Laboratory**

International Partner: **None**

Launch Location: **Kennedy Space Center, FL**

Launch Vehicle: **TBD**

Mission Duration: **3 year science mission**

Requirement Derived from: **2011 Planetary Science Decadal Survey**

Budget Portfolio: **Science, Planetary Science**

PROJECT SUMMARY

The Europa Clipper project plans to hold its preliminary design review in August 2018 and its confirmation review in October 2018 on schedule, at which point the project will establish its cost and schedule baselines. The project implemented a process to manage instrument costs with options to descope an instrument entirely or in part but has not descoped any instruments to date. The project continues to work toward a June 2022 launch but the availability of a qualified and well-understood launch vehicle could affect that schedule. The project is maintaining compatibility with multiple launch vehicles as it awaits NASA's selection. The project is evaluating design options that may affect the mass and power needed to operate the spacecraft and its 10 instruments. The project has experienced workforce shortages since October 2016, but has seen recent improvements. The project is currently holding cost reserves consistent with Jet Propulsion Laboratory (JPL) policy, but is not meeting JPL's schedule reserve requirement due to technical issues with solar array development.

PROJECT BUDGET INFORMATION
percent of current portfolio for fiscal year 2019

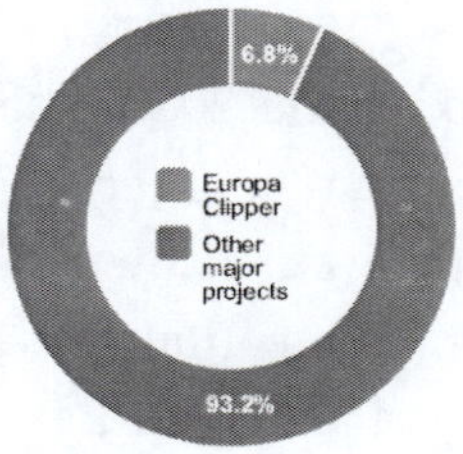

TOTAL FUNDED TO DATE
then-year dollars in millions

PRELIMINARY COST[a]
then-year dollars in millions

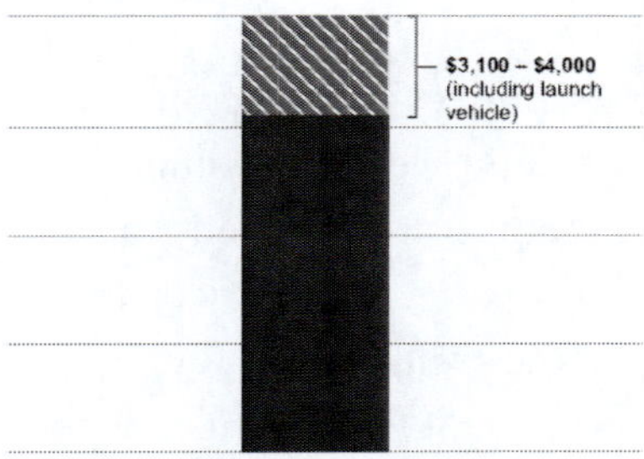

[a]*This estimate is preliminary, as the project is in formulation and there is uncertainty regarding the costs associated with the design options being explored. NASA uses these estimates for planning purposes.*

PRELIMINARY SCHEDULE

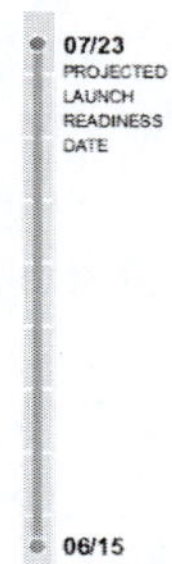

LATEST ESTIMATE
FEB. 2018

Cost and Schedule Status

In February 2017, NASA approved the Europa Clipper project to proceed with preliminary design and technology completion, but the project's independent review board highlighted that the complexity and scope of the current 10 instruments may cause cost and schedule growth.

In response, the project proposed and NASA approved a process to manage instrument costs with options to descope an instrument entirely or in part to prevent cost growth. For example, if an instrument exceeds its development cost by 20 percent, the project would propose a descope option to NASA that brings instrument cost below that threshold. NASA has not descoped any instruments to date.

The project is currently holding cost reserves consistent with JPL policy, but is not meeting JPL's schedule reserve requirement due to technical issues with solar array development. In March 2017, the project discovered that the materials it planned to use on its solar arrays did not meet requirements under low temperatures. As a result, the project changed the materials it plans to use and is exploring using a different vendor for the solar arrays. Project officials said the technical issues and the project's actions to mitigate them delayed delivery of the solar arrays 3 months beyond the project's baseline need date. However, project officials said the baseline need date was earlier than the actual need date for solar array integration. As a result, the project is reassessing the integration and test flow and may be able to recover some schedule reserves.

The project plans to hold its preliminary design review in August 2018 and its confirmation review in October 2018 on schedule, at which point the project will establish its cost and schedule baseline.

Launch

The Europa Clipper project continues to work toward a June 2022 launch, but the availability of a qualified and well-understood launch vehicle could affect that schedule. Project officials stated that the project needs to select a launch vehicle by the end of 2018 to maintain a 2022 launch date. The project is maintaining compatibility with multiple launch vehicles such as the Delta IV Heavy, Falcon Heavy, and NASA's Space Launch System (SLS); however, each of the launch vehicles under consideration has trade-offs that affect development. The Consolidated Appropriations Act, 2016 requires the project to use SLS; however, recent SLS schedule delays could result in a delay to Europa Clipper's planned June 2022 launch readiness date unless NASA adjusts its current plans for SLS and the associated ground systems. Project officials support using SLS as the launch vehicle, even if it delays the planned launch, because SLS allows a direct flight to Jupiter. This path could reduce the travel time by approximately 4.5 years compared to other launch vehicle options.

Technology and Design

The project is currently meeting JPL mass and power margin requirements, but is evaluating design options that may affect the mass and power needed to operate the spacecraft and its 10 instruments. For example, accommodation of the ice-penetrating radar has been difficult because it requires a higher-than-expected amount of power that could lead to design trades. To mitigate this risk, the project is conducting additional testing of the instrument to evaluate compatibility at lower levels of power. The project is also evaluating design options for its solar arrays, which project officials said will increase available power but may also increase mass.

Other Issues to Be Monitored

The project experienced workforce shortages at JPL in several key areas since October 2016. The project has revised staffing plans several times, but the project continues to face workforce shortages even with NASA headquarters involvement. As of January 2018, project officials reported that improvements have been seen in addressing workforce shortages. Other JPL projects have experienced workforce issues as well, which NASA attributes to the amount of work the laboratory is managing.

Project Office Comments

Europa Clipper project officials provided technical comments on a draft of this assessment, which were incorporated as appropriate.

Exploration Ground Systems

The Exploration Ground Systems (EGS) program is modernizing and upgrading infrastructure at the Kennedy Space Center and developing software needed to integrate, process, and launch the Space Launch System (SLS) and Orion Multi-Purpose Crew Vehicle (Orion). The EGS program consists of several major construction and facilities projects including the Mobile Launcher (pictured to the left), Crawler Transporter, Vehicle Assembly Building, and launch pad, all of which need to be complete before the first uncrewed exploration mission using the SLS and Orion vehicles.

Source: NASA | GAO-18-260SP

FORMULATION
02/12 Formulation start
08/12 System requirements/ mission definition review
03/14 Preliminary design review
05/14 Project confirmation

IMPLEMENTATION
12/15 Critical design review
01/18 GAO review
04/18 System integration review
11/18 Committed launch readiness date
12/19-06/20 Replanned launch readiness date

PROJECT INFORMATION

NASA Lead Center: **Kennedy Space Center**

International Partner: **None**

Requirement Derived from: **NASA Authorization Act of 2010**

Budget Portfolio: **Exploration, Exploration Systems Development**

PROJECT SUMMARY

In December 2017, NASA announced a new schedule and cost estimate for the EGS program, after determining that the November 2018 launch readiness date for Exploration Mission-1 (EM-1) was no longer feasible. NASA is now planning to a launch readiness date of December 2019, with 6 months of schedule reserve to a June 2020 date. The life-cycle cost estimate for EGS is approximately $3.1 billion to the December 2019 date and $3.2 billion to June 2020. The program's schedule is currently driven by its two major software development efforts, both of which are undergoing technical challenges. For example, the ground control system is in the process of transitioning to a new contractor and is undergoing organizational changes. In addition, while the program has made progress with major projects such as the Vehicle Assembly Building and Launch Pad 39B, the Mobile Launcher is continuing to have technical challenges. Fifteen percent of the Mobile Launcher's accessories and umbilicals still have to complete testing, and there are emerging concerns about the structural integrity of the Mobile Launcher's base.

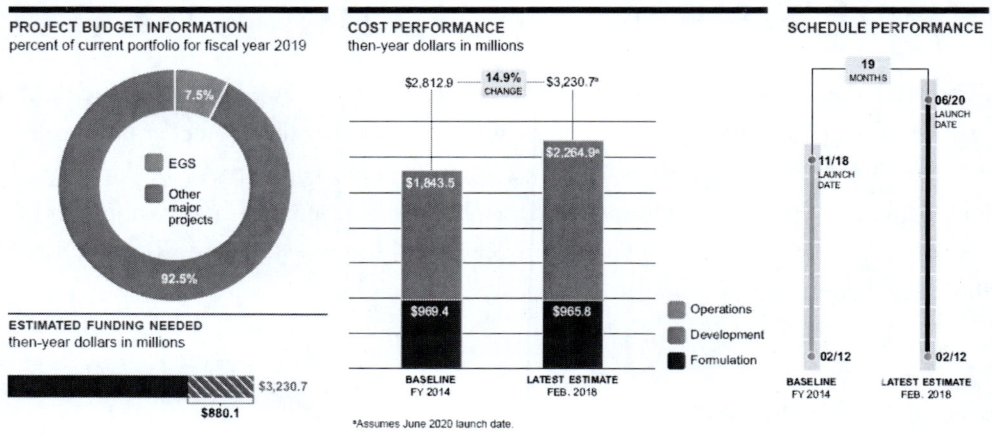

Cost and Schedule Status

In December 2017, NASA announced a new schedule and cost estimate for the EGS program, after determining that the November 2018 launch readiness date for Exploration Mission-1 (EM-1) was no longer feasible. In April 2017, we found that the date was likely unachievable for all three human spaceflight programs—EGS, the Space Launch System, and the Orion Multi-Purpose Crew Vehicle—due to technical challenges continuing to cause schedule delays and the programs having little to no schedule reserve to the EM-1 date, meaning they would have to complete all remaining work with little margin for error for unexpected challenges that could arise.[45] The new launch readiness date for EM-1 is now December 2019, but NASA has allocated 6 months of schedule reserve to June 2020 for possible manufacturing and production schedule risks. This represents a delay of 3-19 months for EM-1. The life-cycle cost estimate for EGS is approximately $3.1 billion to the December 2019 date and $3.2 billion to June 2020, or 11.8 to 14.9 percent above the project's committed baseline.

Technology

EGS's two major software development efforts—Spaceport Command and Control System (SCCS), which will operate and monitor ground equipment, and Ground Flight Application Software (GFAS), which will interface with flight systems and ground

[45] GAO-17-414.

crews—are the program's primary schedule drivers, and the program has recently restructured the organization of the SCCS in order to address gaps in systems engineering processes, among other concerns. EGS's other major software development effort, GFAS, continues to face challenges due to late deliveries from the SLS and Orion programs and delays in SCCS; however, program officials stated they have been heavily focused on resolving SCCS challenges. According to program officials, increased processing needs combined with unplanned rework led to higher complexity in the SCCS system. This complexity in turn led to a program review of SCCS, which found gaps in the system development process and in SCCS's organizational structures. As a result of these issues, EGS is restructuring the Systems Engineering and Integration organization, augmenting hardware, and hiring more staff. Program officials noted that a challenge in this restructuring is hiring because NASA is competing for software developers with the local aerospace economy. The program is also executing a pre-planned transition that moves SCCS under the same contractor as GFAS and will shift the SCCS development structure to smaller deliverables that can be released more often. A senior NASA official said that EGS's software development is a top issue leading up to EM-1, and that as a result, the EGS Associate Program Manager was temporarily reassigned to exclusively manage enhancements to the EGS software development effort.

Design

The EGS program has made progress on several projects, including the Vehicle Assembly Building and Launch Pad 39B, but technical challenges continue with the Mobile Launcher. The Mobile Launcher supports the assembly, testing, and servicing of the SLS rocket and provides the platform on which SLS and Orion will launch. The program has started verification and validation activities for the Mobile Launcher, while ground support equipment and umbilical installation projects are still underway. For example, 15 percent of the Mobile Launcher's umbilicals and launch accessories still have to complete testing. Program officials previously said this concurrency increases risk because of uncertainties regarding how systems not yet installed may affect the systems already installed. In November 2017, EGS reported that loads models have indicated low stress margins in critical locations in the Mobile Launcher base. The program attributed this issue to an error in their model. The program is currently designing structural reinforcements for the base of the Mobile Launcher and hydraulics pedestals for the launch pad. Program officials stated this approach could be completed within the program's existing schedule.

Project Office Comments

In commenting on a draft of this assessment, EGS program officials said the vast majority of EGS development and production content is on track for EM-1 and work is proceeding for the first flight of crew on Exploration Mission-2. In addition, officials said EGS has strengthened its ground software development plans to reflect expected deliveries from other programs. Program officials also provided technical comments, which were incorporated as appropriate.

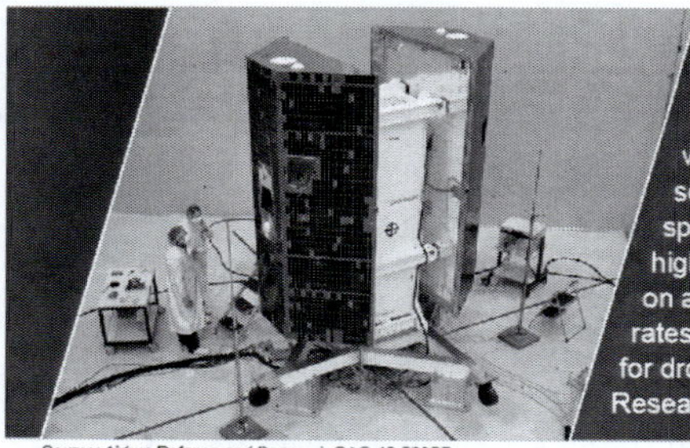

Gravity Recovery and Climate Experiment Follow-On

The Gravity Recovery and Climate Experiment Follow-On (GRACE-FO) will continue and expand upon the 2002 GRACE mission, which ended science operations in October 2017. The system, which consists of two spacecraft working together to obtain scientific measurements, will provide high-resolution models of Earth's gravity field and insight into water movement on and beneath the Earth's surface for up to 5 years. These models will provide rates of ground water depletion and polar ice melt and enable improved planning for droughts and floods. GRACE-FO is a collaborative effort with the German Research Centre for Geosciences (GFZ).

Source: Airbus Defense and Space. | GAO-18-280SP

FORMULATION
- 01/12 Formulation start
- 07/12 System requirements/ mission definition review
- 01/14 Preliminary design review
- 02/14 Project confirmation

IMPLEMENTATION
- 02/15 Critical design review
- 07/15 System integration review
- **01/18 GAO review**
- 02/18 Committed launch readiness date
- 05/18 Manifested launch readiness date

PROJECT INFORMATION

NASA Lead Center: **Jet Propulsion Laboratory**

International Partner: **German Research Centre for Geosciences (GFZ)**

Launch Location: **Vandenberg Air Force Base, CA**

Launch Vehicle: **Falcon 9**

Mission Duration: **5 years**

Requirement Derived from: **NASA 2010 Climate Plan**

Budget Portfolio: **Science, Earth Science**

PROJECT SUMMARY

The GRACE-FO project plans to launch in May 2018, 3 months later than its committed launch readiness date of February 2018, but within its committed cost baseline. Delays to the project's launch readiness date, which over time have accumulated to 9 months of delays, are related to the project's launch vehicle arrangement. First, GFZ—NASA's partner on the project, which is responsible for the launch vehicle—reported that it was notified by the Russian Federal Space Agency that the Dnepr launch vehicle was no longer available. Subsequently, GFZ arranged to launch the two GRACE-FO spacecraft, along with commercial satellites, on a SpaceX Falcon 9. But the project experienced subsequent delays after SpaceX experienced an anomaly on a launch pad that resulted in delays to the launch schedule. The project used the extra time created by the launch delays to conduct additional risk reduction activities during system-level integration and test.

PROJECT BUDGET INFORMATION
percent of current portfolio for fiscal year 2019

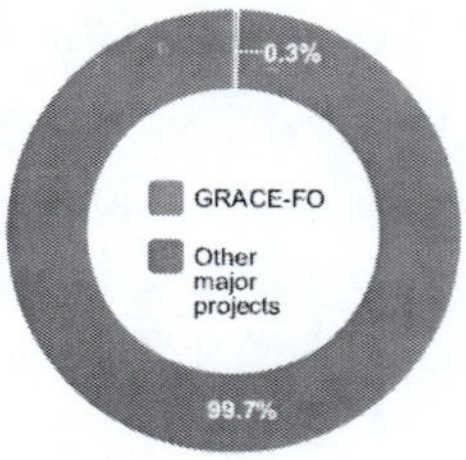

ESTIMATED FUNDING NEEDED
then-year dollars in millions

COST PERFORMANCE
then-year dollars in millions

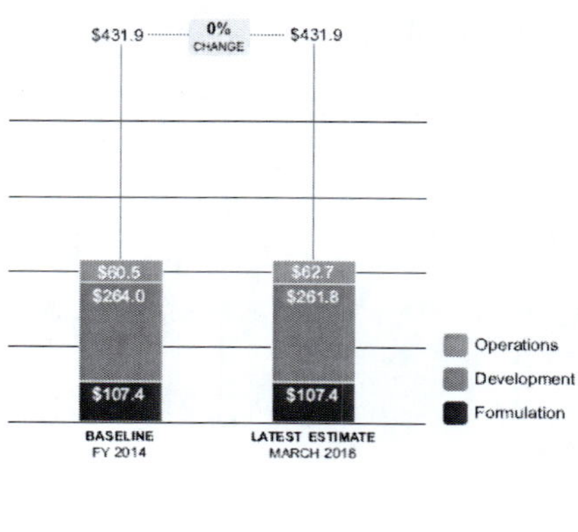

SCHEDULE PERFORMANCE

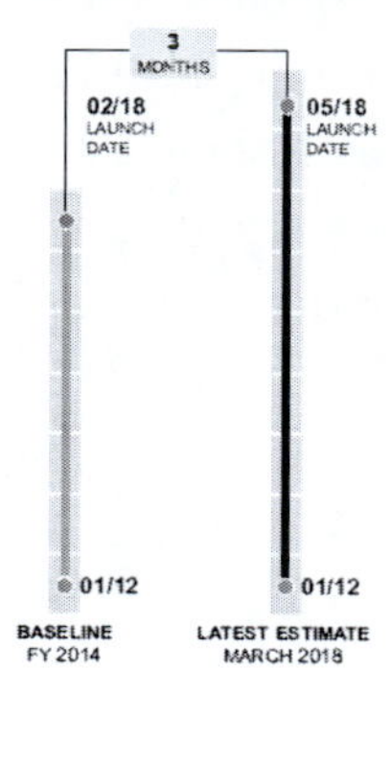

Cost and Schedule Status

The GRACE-FO project plans to launch in May 2018, 3 months later than its committed launch readiness date of February 2018, due to delays related to its launch vehicle arrangement. The project expects to launch within its cost baseline and plans to use its remaining project-held reserves to cover costs through May 2018. The project completed system-level testing on both spacecraft in November 2017 and shipped GRACE-FO to Vandenberg Air Force Base in California for pre-launch activities in December 2017.

Launch and Development Partner

The GRACE-FO project previously planned to launch early in August 2017, but delays related to its launch vehicle arrangement resulted in a 9-month delay to its launch readiness date. The launch vehicle is the responsibility of NASA's partner on the project—GFZ. An initial delay of 4 months occurred when the project changed its launch vehicle after GFZ reported that it was notified by the Russian Federal Space Agency that the Dnepr launch vehicle was no longer available. GFZ, in June 2016, arranged to launch the two GRACE-FO spacecraft, along with commercial satellites, on a SpaceX Falcon 9. SpaceX then experienced an anomaly on a launch pad in September 2016 that resulted in delays to the launch schedule as SpaceX investigated the source of the anomaly. In July 2017, a new 30-day launch readiness window from March to April 2018 was negotiated, resulting in an additional 4 months of delays for the project. In March 2018, the launch was delayed an additional month to May 2018 due to launch range availability.

Integration and Test

The project used the extra time in the schedule due to launch delays to conduct additional risk reduction activities during system-level integration and test. For example, the project identified a design issue with its global positioning system antenna during testing. With the extra time, it replaced the antennas on both spacecraft to resolve the issue. In addition, the project identified a manufacturing issue with one of the two cavities that stabilizes the laser of its Laser Ranging Interferometer (LRI), which is contributed by GFZ. LRI is a technology demonstration that performs the same ranging measurements of GRACE- FO's microwave instrument but with greater precision and is not needed to meet the mission's science requirements. With the extra time, the project was able to replace and reintegrate the cavity with the spacecraft.

In addition, the project is investigating the shifting of an optical system component of LRI during system-level environmental testing in August 2017. The project is investigating the cause, but anticipates the shift is likely due to the expansion of adhesive bonds between the optical components when exposed to temperature and humidity. The project conducted additional tests using spare flight hardware in a vacuum, which removes humidity, to understand future alignment performance in a space-like environment. The project expects that it will be able to use the component as-is because tests indicate there is sufficient performance margin even with the shifting, that exposure to the vacuum of space will reduce the shifting, and it can make operational adjustments to further improve performance, if needed.

Project Office Comments

GRACE-FO project officials provided technical comments on a draft of this assessment, which were incorporated as appropriate.

Ice, Cloud, and Land Elevation Satellite-2

The Ice, Cloud, and Land Elevation Satellite-2 (ICESat-2) is a follow-on mission to ICESat that will measure changes in polar ice-sheet mass and elevation. The measurements will provide researchers a better understanding of the mechanisms that drive polar ice changes and their effect on global sea level. ICESat-2's upgraded laser instrument will allow the satellite to make more frequent measurements and provide better elevation estimates over certain types of terrain than ICESat.

Source: NASA. | GAO-18-280SP

FORMULATION
- 12/09 Formulation start
- 05/11 System requirements/ mission definition review
- 10/12 Preliminary design review
- 12/12 Project confirmation

IMPLEMENTATION
- 02/14 Critical design review
- 05/14 Project rebaseline
- 08/17 System integration review
- **01/18 GAO review**
- 06/18 Committed launch readiness date
- 09/18-10/18 Manifested launch readiness date

PROJECT INFORMATION

NASA Lead Center: **Goddard Space Flight Center**

International Partner: **None**

Launch Location: **Vandenberg Air Force Base, CA**

Launch Vehicle: **Delta II**

Mission Duration: **3 years**

Requirement Derived from: **2007 Earth Science Decadal Survey**

Budget Portfolio: **Science, Earth Science**

PROJECT SUMMARY

The project will miss its rebaselined committed launch date but does not expect the cost to exceed the rebaseline because it has sufficient cost reserves to cover the delay. The project will miss its committed launch readiness date of June 2018 due to issues with its sole instrument—the Advanced Topographic Laser Altimeter System (ATLAS). In July 2016, the ICESat-2 project encountered problems with the flight lasers developed for ATLAS that caused NASA to delay the launch date to October 2018, which is 17 months beyond the original baseline and 4 months beyond the rebaselined date set in 2014. The project held its system integration review and began system-level integration and test in August 2017. The two repaired lasers that will fly on the ATLAS instrument were shipped back to Goddard Space Flight Center, completed flight acceptance testing, and are integrated onto the instrument. The project entered the system assembly and testing phase in October 2017.

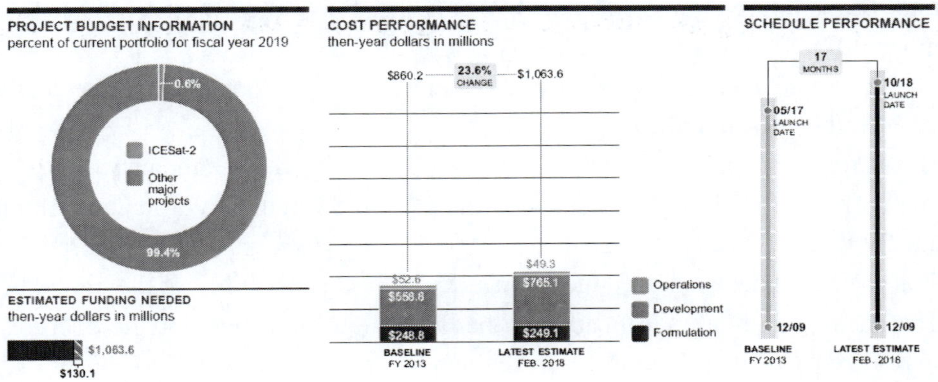

Cost and Schedule Status

The project will miss its rebaselined committed launch date but does not expect the cost to exceed the rebaseline because it has sufficient cost reserves to cover the delay. The ICESat-2 project is proceeding to integration and test after problems with lasers within the primary instrument—the Advanced Topographic Laser Altimeter System (ATLAS)—caused NASA to delay the launch readiness date from June 2018 to October 2018. This represents a 17-month delay beyond the original baseline and a 4-month delay beyond the rebaselined date set in 2014. However, the project is working to an earlier, September 2018 launch readiness date. The project completed a cost and schedule replan in October 2017, which required a large portion of remaining headquarters-held cost reserves, but did not take the project over its baseline commitment of $1,063.5 million.[46] The project is holding cost and schedule reserves at the level required by NASA center policy, but the project will need to realign funding planned for system integration and testing to address a shortfall in cost reserves for the operations phase. The project held its system integration review in August 2017 and received approval to enter the system assembly and testing phase in October 2017.

Technology and Design

Over numerous years, the ICESat-2 project has addressed issues with its flight lasers, and continues to mitigate the risk that they might fail again in the future. During ATLAS environmental testing in 2016, the project determined two of its three lasers needed repairs due to cracked crystal. The remaining laser did not require rework and continued to

[46] This total baseline cost differs from the total cost outlined in the table above due to an increase in project labor during the formulation phase. This increase is not related to the ATLAS instrument.

operate—completing over 2,300 hours of runtime. Subsequently, the project found that components on a test model—similar to the laser that did not need repairs—cracked while in storage. Project officials were unable to identify a single root cause of the cracking but determined this was an indication that the unrepaired laser could fail. ICESat-2 only needs one laser for mission success, but will carry a second for redundancy. In July 2017, following the completion of a trade study to determine which two of its three flight lasers should fly on ATLAS, NASA decided to use the two lasers that were repaired for flight. Officials said this decision was due to uncertainty in the reliability of the unrepaired laser and that the repaired lasers fully address the factors that contributed to the fractures. Since being repaired, together these lasers have completed a total of over 1,200 hours of runtime and have performed to requirements. Flying the two repaired lasers does not affect the project's plans to launch in September 2018. However, the project is carrying multiple risks to account for the potential that the repaired lasers could still fail at various points in time. For example, the project is carrying two risks that weigh the effect of the lasers failing on the ground during integration and test or failing while on orbit. If a laser fails during integration and test, it would take 5 weeks to replace it with a spare.

Integration and Test

The ICESat-2 project is proceeding through integration and test now that the ATLAS flight lasers are integrated onto the instrument and undergoing performance testing. The project held its system integration review and began system-level integration and test in August 2017. The two repaired lasers that will fly on the ATLAS instrument were shipped back to Goddard Space Flight Center and completed flight acceptance testing. On one of the repaired lasers, which was fully integrated onto the optical bench, the project observed an unusual pattern in its optical components during initial performance testing, and ATLAS officials have convened a review board to investigate the cause of this issue. The other repaired laser, which performed as expected during acceptance testing, began integration in September 2017 and is now undergoing performance testing.

Project Office Comments

In commenting on a draft of this assessment, ICESat-2 officials stated that during initial ATLAS testing, flight laser two exhibited an anomalous laser behavior. Project officials stated the laser was transported back to the contractor for cleaning and testing, which revealed comparable results to previous tests. The laser was returned to NASA and is performing normally. Project officials also provided technical comments, which were incorporated as appropriate.

Interior Exploration using Seismic Investigations, Geodesy, and Heat Transport

The Interior Exploration using Seismic Investigations, Geodesy, and Heat Transport (InSight) is a Mars lander with two primary objectives. It is intended to further understanding of the formation and evolution of terrestrial planets by determining Mars's size, its composition, and the physical state of the core; the thickness of the crust; and the composition and structure of the mantle, as well as the thermal state of the interior. It will also determine the present level of tectonic activity and the meteorite impact rate on Mars. InSight is based on the Phoenix lander design. Phoenix successfully landed on Mars in 2008.

Source: © California Institute of Technology | GAO-18-280SP

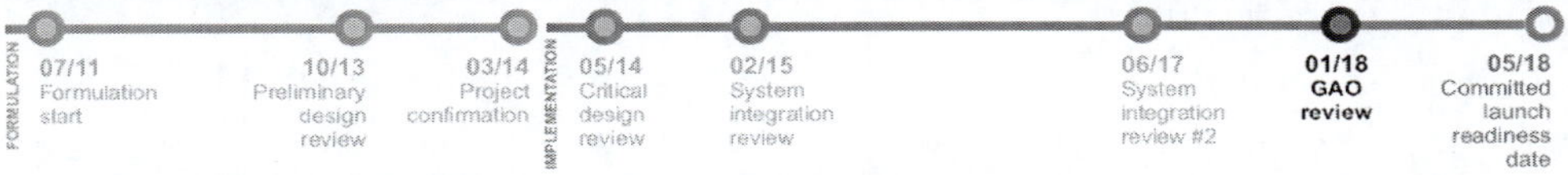

PROJECT INFORMATION

NASA Lead Center: **Jet Propulsion Laboratory**

International Partners: **Centre National d'Etudes Spatiales (France) and German Aerospace Center (Germany)**

Launch Location: **Vandenberg Air Force Base, CA**

Launch Vehicle: **Atlas V**

Mission Duration: **29 months**

Requirement Derived from: **2011 Planetary Science Decadal Survey**

Budget Portfolio: **Science, Planetary Science**

PROJECT SUMMARY

The InSight project is on track to meet its revised cost and schedule agreement after technical issues with its primary instrument—the Seismic Experiment for Interior Structure (SEIS)—caused it to miss the previously planned launch date of March 2016. The project is holding cost and schedule reserves consistent with the level required by Jet Propulsion Laboratory policy. NASA redesigned the SEIS container and integrated it on the lander in August 2017. The project continues to track a risk that sensors within the SEIS will not function as intended, which would affect the project's ability to meet science requirements. To increase confidence that the sensors will perform, the project is completing tests to confirm the sensors' ability to survive for the mission's duration. The project also used the additional time from the launch delay to redesign the wiring within the HP3—a German-contributed instrument that drills into the Martian surface to collect temperature data—to address an anomaly found in testing.

PROJECT BUDGET INFORMATION
percent of current portfolio for fiscal year 2019

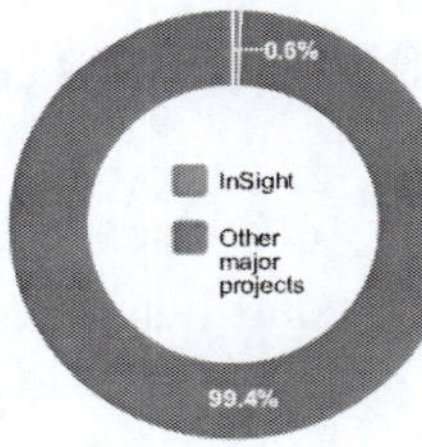

ESTIMATED FUNDING NEEDED
then-year dollars in millions

COST PERFORMANCE
then-year dollars in millions

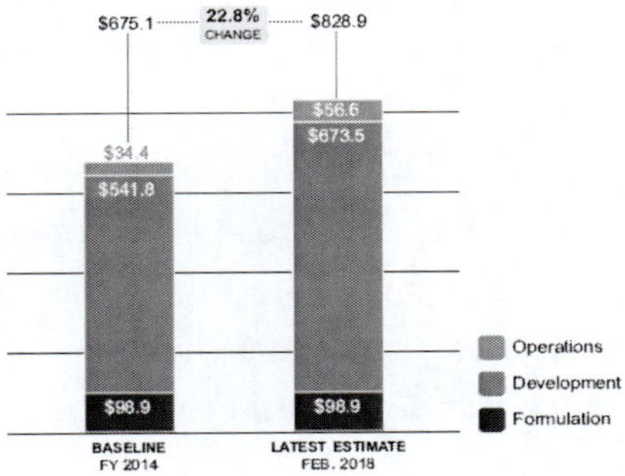

SCHEDULE PERFORMANCE

Cost and Schedule Status

The InSight project is on track to meet its revised cost and schedule agreement approved by NASA in August 2016 after the project missed its previously planned launch date. The InSight project missed its original launch date due to technical issues with the Seismic Experiment for Interior Structure (SEIS) instrument—the project's primary instrument and a contribution from the Centre National d'Etudes Spatiales (CNES)—which increased costs by almost $154 million and delayed the launch by 26 months. The project delivered the InSight spacecraft to the launch site in February 2018 before its planned launch in May 2018. The project is holding cost and schedule reserves consistent with the level required by Jet Propulsion Laboratory policy.

Integration and Test and Development Partner

The InSight project has resolved technical issues that necessitated a SEIS redesign, but is working to mitigate a risk on a key SEIS component that could affect SEIS's performance in operations. After discovering vacuum seal leaks that prevented the SEIS container from sealing properly and ultimately caused the launch delay, the project restructured SEIS responsibilities with NASA redesigning the SEIS container and CNES conducting integration and test. In March 2017, CNES confirmed that the redesign of the SEIS container was successful after completing a leak test, and integrated the SEIS with the lander in August 2017. The project is tracking a risk related to a potential failure of a Very Broad Band (VBB) sensor—a key SEIS component that will be used to measure seismic activity— in operations. The project has requested and received a significant number of VBB waivers. For example, the project requested and received approval to proceed with an oxidized part on several VBBs. The waivers individually have a low risk of causing a VBB failure, but together present a more moderate risk of a failure. More than one VBB failure would prevent the project from meeting its science requirements, according to officials. The project is completing tests to confirm the sensors' ability to survive for the mission's duration, which should provide increased confidence in the VBBs.

Because the project had additional time in its schedule due to the launch delay, it redesigned the Heat Flow and Physical Properties Probe (HP3)—a contributed instrument that will drill into the Martian surface to take temperature measurements—to resolve anomalies that occurred late in testing and to improve performance. These anomalies included damaged internal wiring caused by shocks generated by the hammering mechanism. The project completed a HP3 trade study, which evaluated whether or not to fly the redesigned HP3 or proceed with the original design, and determined the redesigned HP3 represented the lower risk option for the mission. The project integrated the redesigned HP3 in October 2017 prior to system-level testing. The HP3 is not needed for the project to meet its level 1 science requirements.

Design

The InSight project was tracking a risk related to its parachute that could have consumed approximately 27 percent of its cost reserves, but in October 2017 the project made the decision to use the existing parachute design that removed that risk. According to officials, the InSight project uses the same manufacturer for nylon in its parachute as the Mars 2020 project—another project that will land on Mars—originally planned to use. Mars 2020 discovered a potential problem with the strength of its parachute fabric as the fabric was degrading upon exposure to high heat during a planetary protection activity. As a result, the InSight project was concurrently building a parachute with new nylon material as well as a parachute using existing material. However, the project decided to abandon building a parachute using new material when it was determined that the new nylon material did not meet specifications, the vendor could not meet the required schedule in order to make the risk reduction option viable, and, according to officials, expert reviews rated the existing parachute as acceptable to fly. The Advanced Supersonic Parachute Inflation Research Experiment successfully completed a test in October 2017 that provided the project with additional support for their decision to use the parachute using original material.

Project Office Comments

InSight project officials provided technical comments on a draft of this assessment, which were incorporated as appropriate.

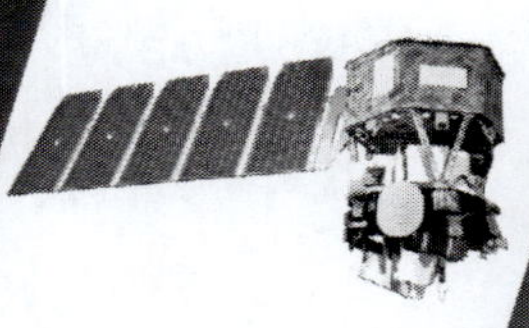

Ionospheric Connection Explorer

The Ionospheric Connection Explorer (ICON) observatory will orbit Earth to explore its ionosphere—the boundary region between Earth and space where ionized plasma and neutral gas collide and react. Its four instruments will make direct measurements and use remote sensing to further researchers' understanding of Earth's upper atmosphere, the Earth-Sun connection, and the ways in which Earth weather drives space weather.

Source: University of California, Berkeley. | GAO-18-280SP

FORMULATION
- 09/11 Formulation start
- 01/14 System requirements/ mission definition review
- 07/14 Preliminary design review
- 10/14 Project confirmation

IMPLEMENTATION
- 04/15 Critical design review
- 08/16 System integration review
- 10/17 Committed launch readiness date
- **01/18 GAO review**
- **No earlier than 06/18** Manifested launch readiness date

PROJECT INFORMATION

NASA Lead Center: **Goddard Space Flight Center**

International Partner: **Centre Spatial de Liege Université de Liège (Belgium)**

Launch Location: **Kwajalein (Marshall Islands)**

Launch Vehicle: **Pegasus**

Mission Duration: **2 years**

Requirement Derived from: **2010 Science Mission Directorate Science Plan and 2009 Heliophysics Roadmap Team Report to the NASA Advisory Council**

Budget Portfolio: **Science, Heliophysics**

PROJECT SUMMARY

The ICON project has missed its committed launch readiness date of October 2017 and both the project's schedule and costs are under review. Delays to the project's launch readiness date are related to problems with ICON's launch vehicle. The project experienced a 6-month delay due to a transport incident with a launch vehicle segment and launch range conflicts. In September 2017, an anomaly found in testing of the launch vehicle bolt cutter assemblies resulted in additional delays, but the magnitude of these delays is unknown while an investigation of the anomaly is underway. In February 2018, NASA determined the project will launch no earlier than June 2018, but this date is still under review. The ICON project completed its system-level integration and test activities in April 2017 as planned, but has not shipped the spacecraft to the launch site due to the launch delay.

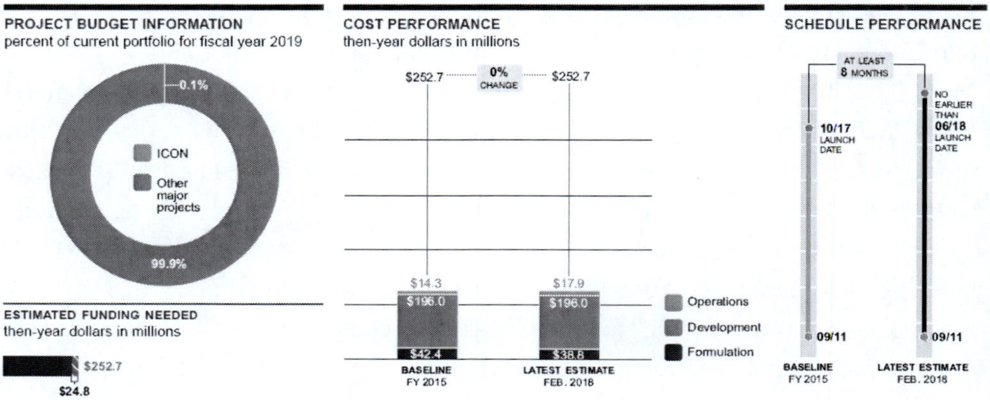

Cost and Schedule Status

The ICON project has missed its committed launch readiness date of October 2017 and both the project's schedule and costs are under review. The project completed its system-level integration and test activities in April 2017 as planned, but has not shipped the spacecraft to the launch site due to ongoing delays related to the launch vehicle. According to officials, the project received $7.8 million of headquarters-held reserves to cover initial delays, but it cannot determine whether it will exceed its cost baseline until NASA and the launch vehicle provider set a new launch date. Project officials said they need about $1 million a month for ICON to remain in storage.

Launch

The ICON project had planned to launch early, in June 2017, but the project has experienced delays associated with its launch vehicle. In January 2017, two of the Pegasus launch vehicle's three stages were involved in a transport accident. The stages were subsequently returned to the launch vehicle contractor facility for inspection and testing, and no damage was found. The launch vehicle contractor then delivered the stages to Vandenberg Air Force Base for integration and testing activities. Due to conflicts at the launch vehicle range, the earliest available launch date was December 2017, which resulted in a 6-month launch delay from the planned June 2017 launch date.

In September 2017, however, an anomaly identified in bolt cutter assembly confidence testing—testing to show that the bolts that hold the launch vehicle and payload together will separate as planned during launch—resulted in additional delays, but the magnitude of the delay is unknown. One of nine bolt cutter assemblies failed to fracture a bolt during testing. As a result, NASA and the contractor halted testing and began an investigation of the anomaly, which is ongoing. NASA's Launch Services Program is working with the launch vehicle provider to identify the root cause of the anomaly, evaluate options to resolve the issue, and determine a new launch readiness date. In February 2018, NASA determined the project will launch no earlier than June 2018, but this date is still under review.

Project Office Comments

In commenting on a draft of this assessment, ICON project officials stated that the project completed all observatory integration and test activities last year. As of January 2018, the observatory remains in the Orbital-ATK cleanroom in Gilbert, Arizona in a safe state—under continuous purge and performing periodic monitoring of the battery voltage—awaiting determination of a new launch date and shipment for launch vehicle integration. Project officials also provided technical comments, which were incorporated as appropriate.

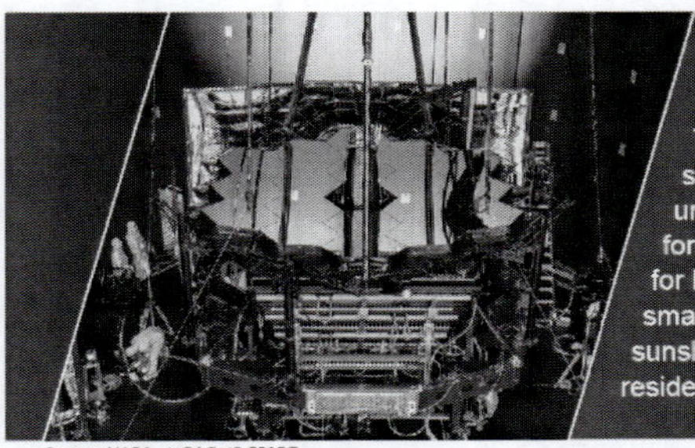

James Webb Space Telescope

The James Webb Space Telescope (JWST) is a large, infrared-optimized space telescope designed to help understand the origin and destiny of the universe, the creation and evolution of the first stars and galaxies, and the formation of stars and planetary systems. It will also help further the search for Earth-like planets. JWST will have a large primary mirror composed of 18 smaller mirrors and a sunshield the size of a tennis court. Both the mirror and sunshield are folded for launch and open once JWST is in space. JWST will reside in an orbit about 1 million miles from the Earth.

Source: NASA. | GAO-18-280SP

FORMULATION

- 03/99 Formulation start
- 01/06 Mission/system definition review
- 03/08 Preliminary design review
- 07/08 Project confirmation

IMPLEMENTATION

- 03/10 Critical design review
- **01/18 GAO review**
- 06/18 System integration review
- 10/18 Committed launch readiness date
- 05/20 Target launch readiness date

PROJECT INFORMATION

NASA Lead Center: **Goddard Space Flight Center**

International Partners: **European Space Agency, Canadian Space Agency**

Launch Location: **Kourou, French Guiana**

Launch Vehicle: **Ariane 5**

Mission Duration: **5 years (10-year goal)**

Requirement Derived from: **2001 Astrophysics Decadal Survey**

Budget Portfolio: **Science, Astrophysics**

PROJECT SUMMARY

The JWST project delayed its planned launch readiness date by a total of at least 19 months from October 2018 to May 2020. In March 2018, NASA announced the most recent delay to approximately May 2020 based on a schedule analysis by its standing review board. This analysis showed that more time will be needed to integrate and test the telescope and spacecraft elements and conduct environmental testing. For several years, the observatory contractor has overestimated workforce reductions at the beginning of each fiscal year, but technical challenges have prevented the planned reductions and hindered JWST's ability to control costs. As a result, the observatory contractor has maintained higher workforce levels than expected, and may continue to do so in the coming months. Along with the latest launch delay, this puts the project at risk of exceeding its $8 billion congressional cost cap.

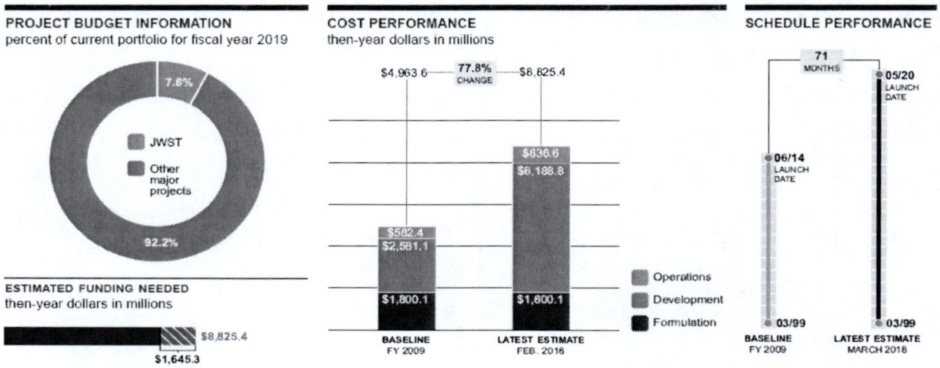

Cost and Schedule Status

The JWST project has delayed its planned launch readiness date by at least 19 months, from the committed October 2018 launch readiness date to approximately May 2020. NASA announced two delays for the project since we last reported in May 2017.a First, in September 2017, the project announced a delay of up to 8 months based on the results of a schedule risk assessment that showed various components of spacecraft element integration were taking longer to complete than expected. Prior to establishing the new launch window, the project used all remaining schedule reserves to address various technical issues, including a test anomaly on the telescope and sunshield hardware challenges. The new launch date included up to 4 months of new schedule reserves. However, by February 2018, the project had consumed the 4 months of reserves due to continuing challenges with spacecraft integration and test. Then, in March 2018, the JWST project announced an 11- month launch readiness delay to approximately May 2020 based on the results of a standing review board schedule analysis. According to the analysis, the previous launch date was not possible due to lessons learned during spacecraft element integration and test and propulsion system rework. The analysis further indicated that more time will be needed to integrate and test the telescope and spacecraft elements and conduct environmental testing.

Due to these challenges, the observatory contractor has maintained higher levels of workforce than expected, and may continue to do so in the coming months. As a result, the project is at risk of exceeding the $8 billion congressional cap for formulation and development costs established in 2011. An external independent review board will conduct a schedule analysis, with the results expected in May 2018, to be followed by the agency's final decision on JWST's launch date in June 2018 and a report to Congress in summer 2018. After the launch date is determined, NASA will update the project's cost estimate.

Integration and Test

The project and observatory contractor significantly underestimated the time required to complete integration and test work on the spacecraft element. Execution of spacecraft integration and test tasks was much slower than planned due to a variety of challenges including complexity of work and reach and access limitations on flight hardware. In addition, the observatory contractor has consumed several weeks of schedule reserves due to various workmanship errors, particularly with respect to the spacecraft propulsion systems. For example, an observatory contractor technician applied too much voltage and damaged components of the propulsion system, and reattaching the replacement components consumed 5 weeks of reserves. Also in May 2017, the observatory contractor discovered that valves in the thruster modules— which help control spacecraft on-orbit positioning—had been damaged by a cleaning solution and had to be refurbished. Reattaching the refurbished modules was expected to be complete by February 2018, but was delayed by one month when a technician applied too much voltage to one of the components in a recently refurbished thruster module.

Contractor

The observatory contractor continued to maintain higher than planned workforce levels in the past year, and may continue to do so in the coming months. For several years, the observatory contractor has overestimated workforce reductions at the beginning of each fiscal year, but technical challenges have prevented the planned reductions and hindered JWST's ability to control costs.

After completing negotiations on a cost overrun proposal in September 2017, driven by higher-than-planned workforce levels, the project is expected to issue a request for proposal from the observatory contractor in early 2018 for the costs for the remaining work through the new launch window. A cost overrun proposal seeks to increase the value of a cost-reimbursement contract when the total estimated cost is less than the contract's estimated cost to complete the performance of the contract.

Project Office Comments

JWST project officials provided technical comments on a draft of this assessment, which were incorporated as appropriate.

Source: Orbital ATK (artist rendering) | GAO-18-280SP

Landsat 9

Landsat 9 is the next satellite in the Landsat series Program, which provides a continuous space-based record of land surface observations to study, predict, and understand the consequences of land surface dynamics, such as deforestation. The program is a collaborative, joint mission between NASA and the U.S. Geological Survey. The Landsat data archive constitutes the longest continuous moderate-resolution record of the global land surface as viewed from space and is used by many fields, such as agriculture, mapping, forestry, and geology.

FORMULATION
- 03/15 Formulation start
- 06/16 System requirements/ mission definition review
- 09/17 Preliminary design review
- 12/17 Project confirmation
- **01/18 GAO review**

IMPLEMENTATION
- 04/18 Critical design review
- 08/19 System integration review
- 12/20 Target launch readiness date
- 11/21 Committed launch readiness date

PROJECT INFORMATION

NASA Lead Center: **Goddard Space Flight Center**

International Partner: **None**

Launch Location: **Vandenberg Air Force Base, CA**

Launch Vehicle: **Atlas V**

Mission Duration: **5 years**

Requirement Derived from: **National Plan for Civil Earth Observations**

Budget Portfolio: **Science, Earth Science**

PROJECT SUMMARY

The Landsat 9 project entered the implementation phase in December 2017 and formally established its cost and schedule baselines. The project is working to a launch readiness date of December 2020, 11 months earlier than its baselined launch date, in part to help meet a goal of the scientific community to maintain two Landsat satellites on-orbit simultaneously. As a result, the project is not holding schedule reserves at the level required by NASA center policy to address known and unknown risks. The project considers this schedule challenging, but achievable because of the high use of heritage technologies, mature instrument designs, and extensive use of hardware from the prior Landsat mission, Landsat 8. The project entered implementation with mature technologies and stable instrument designs. The project plans to hold its critical design review in April 2018. Our work on product development best practices shows that at least 90 percent of design drawings should be released by critical design review to lower the risk of subsequent cost and schedule growth. As of January 2018, the project had released about 64 percent of its design drawings. Prior to this review, the project will need to mature the design of its spacecraft.

PROJECT BUDGET INFORMATION
percent of current portfolio for fiscal year 2019

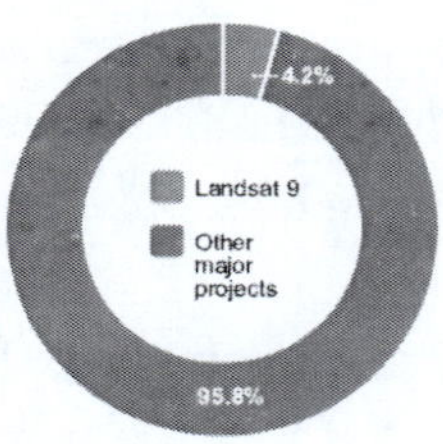

ESTIMATED FUNDING NEEDED
then-year dollars in millions

COST PERFORMANCE
then-year dollars in millions

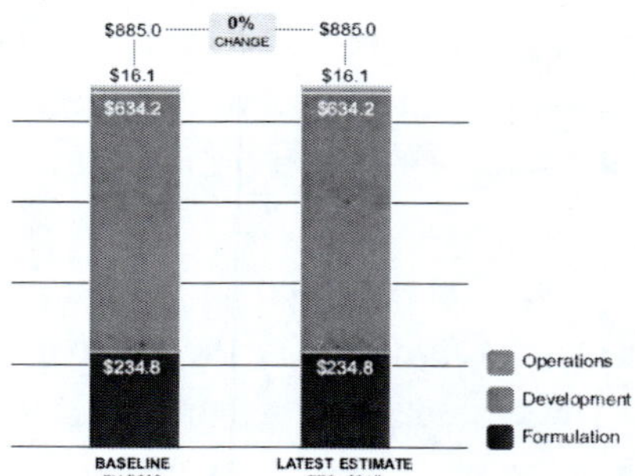

SCHEDULE PERFORMANCE

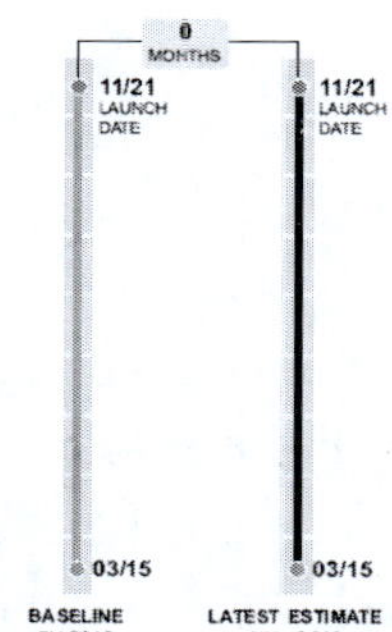

Cost and Schedule Status

The Landsat 9 project entered the implementation phase in December 2017 and formally established its cost and schedule baselines. The project set a baseline life-cycle cost of $885 million and a launch date of November 2021. However, the project is working toward a launch readiness date of December 2020 due to direction in the Explanatory Statement accompanying the Consolidated Appropriations Act, 2016. According to officials, this earlier launch date will help ensure that the Landsat program is able to maintain two Landsat satellites on-orbit simultaneously, which is a goal of the scientific community, but not a requirement for the program. NASA calculated the project's joint cost and schedule confidence level, the likelihood a project will meet its cost and schedule estimate, at 47 percent for the earlier December 2020 launch date. This is slightly lower than the 50 percent required by NASA policy. However, for the baselined launch date of November 2021, the joint cost and schedule confidence level is greater than 70 percent, which meets the requirement in NASA policy.

As a result of working to this earlier launch date, the project is holding schedule reserves below the level required by NASA center policy. This means the project would have less time to address known and unknown risks without affecting its December 2020 launch date. The project considers this schedule challenging, but achievable because of the high use of heritage technologies, mature instrument designs, and extensive use of hardware from the prior Landsat mission, Landsat 8. In addition, project officials said they have confidence that the project can meet this date because they have been executing to this aggressive schedule without needing any schedule reserves. The project is holding cost reserves at the level required by NASA center policy.

Technology and Design

The Landsat 9 project proceeded into implementation with mature technologies. The project held its preliminary design review in September 2017 with mature technologies and stable instrument designs due to high use of heritage technologies and designs from Landsat 8. Utilizing heritage technologies and designs on projects can help to reduce risk and control costs when they are used for similar purposes in similar environments. Specifically, Landsat 9's instruments—Thermal Infrared Sensor-2 and Operational Land Imager-2—are at a technology readiness level 9 nd all components in the flight segments are above a technology readiness level 6. Maturing technologies to a technology readiness level 6 by preliminary design review is a best practice and helps minimize risks for space systems entering product development.

The project also proceeded into implementation with the designs of its two instruments stable. The project plans to hold its critical design review in April 2018. Our work on product development best practices shows that at least 90 percent of design drawings should be released by critical design review to lower the risk of subsequent cost and schedule growth. As of January 2018, the project had released about 64 percent of its design drawings.

The project completed critical design reviews for its two instruments prior to mission preliminary design review due to the high level of heritage designs from Landsat 8, and has released over 90 percent of drawings for both. Most of the drawings that have not been released are for the spacecraft. The project's spacecraft also has a design similar to that of Landsat 8 and uses heritage components from other Earth-orbiting satellites. The project plans to hold its spacecraft critical design review in February 2018.

Project Office Comments

In commenting on a draft of this assessment, Landsat 9 project officials stated that all project elements are successfully making progress and on track to support a December 2020 launch readiness date. Officials also provided technical comments on a draft of this assessment, which were incorporated as appropriate.

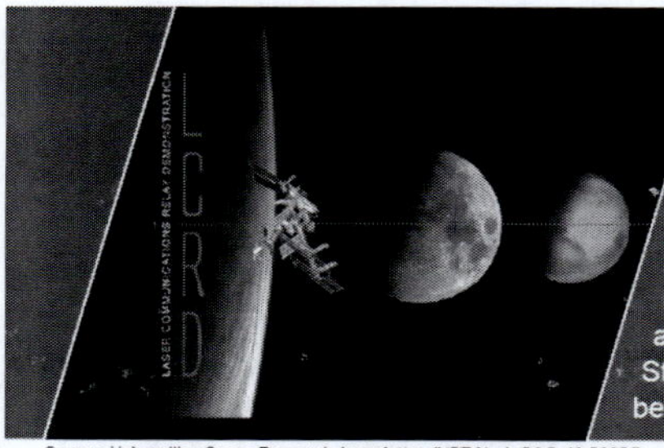

Laser Communications Relay Demonstration

LCRD is a technology demonstration mission with the goal of advancing optical communication technology for use in deep space and near-Earth systems. LCRD will demonstrate bidirectional laser communications between a satellite and ground stations, develop operational procedures, and transfer the technology to industry for future use on commercial and government satellites. NASA anticipates using the technology as a next generation Earth relay as well as to support near-Earth and deep space science, such as the International Space Station and human spaceflight missions. The project is a mission partner and will be a payload on a U.S. Air Force Space Test Program satellite.

Source: Universities Space Research Association (USRA). | GAO-18-280SP

FORMULATION
- 12/12 Formulation start
- 03/13 System requirements/ mission definition review
- 10/13 Preliminary design review
- 12/16 Critical design review
- 04/17 Project confirmation

IMPLEMENTATION
- **01/18 GAO review**
- 07/18 System integration review
- 06/19 Target launch readiness date
- 11/19 Committed launch readiness date

PROJECT INFORMATION

NASA Lead Center: **Goddard Space Flight Center**

International Partners: **N/A**

Launch Location: **Cape Canaveral Air Force Station, FL**

Launch Vehicle: **Atlas V 551**

Mission Duration: **2+ years**

Requirement Derived from: **NASA Strategic Plan**

Budget Portfolio: **Space Technology, Research and Development**

PROJECT SUMMARY

The LCRD project established its cost and schedule baselines in April 2017 at a higher cost and later schedule than preliminary estimates due to funding, programmatic, and design changes made after preliminary design review in October 2013. The LCRD project has had to delay its schedule, change its host spacecraft, and adjust its design to accommodate scope changes as well as funding shortfalls in prior years. LCRD has also added a redundant communication capability that adds longevity but also cost and complexity to the spacecraft. In addition, multiple hardware components have been delivered late and are consuming project schedule reserves heading into integration and test.

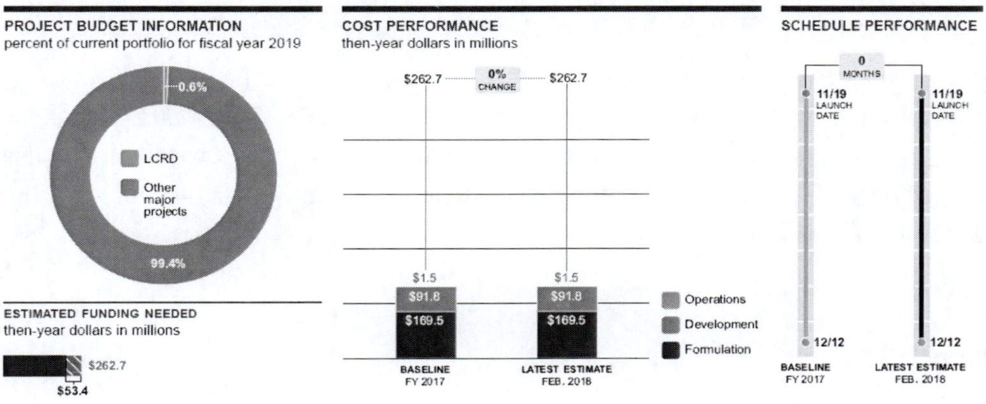

Cost and Schedule Status

The LCRD project's cost and schedule baselines increased compared to its preliminary estimates due to funding, programmatic, and design changes made after preliminary design review in October 2013. LCRD entered the implementation phase in April 2017 and formally established its cost and schedule baselines of $262.7 million to launch by November 2019, which is $23.4 million higher and 23 months later than the preliminary cost estimate. The cost and schedule increases over the preliminary estimate are due in part to externally driven design changes that the project has had to address. The new launch date also reflects the project delaying its confirmation review from late 2013 until April 2017 due to funding and programmatic challenges. For example, funding for fiscal years 2015 and 2016 was less than planned, which led the project to halt development of the ground segment for 2 years. The project has also changed management and undertaken numerous replans and descopes since 2013. As of October 2017, the project was holding cost reserves below levels required by NASA center policy.

Technology

LCRD's technologies were not mature by preliminary design review. As a technology demonstration, Goddard Space Flight Center does not require technologies to be at technology readiness level 6 at that review. However, the same risks of subsequent technical problems apply to technology demonstrations, which can result in cost growth and schedule delays. Further, because technologies are being demonstrated to support future missions, the project added a high-bandwidth radio transceiver as a backup to the optical modules to increase reliability, redundancy, and longevity. A drawback of the laser communication system is that it requires an unobstructed line of sight in order to transmit and receive data. The radio transceiver allows LCRD to transmit data with less interruption as the radio transceiver increases redundancy should the line-of-sight be obstructed, and longevity should an optical unit fail during the mission. In 2017, the proposal for the high-bandwidth radio transceiver came in at a higher than expected cost but LCRD received funding from sources outside the project office. In addition, LCRD is working with its host spacecraft to solve potential on-orbit vibration that could impact data transmission, but the project will have to fund additional damping.

Design

The project held its critical design review in December 2016 with a stable design and then held project confirmation in April 2017, which allowed the project to enter final design and fabrication. The project held its confirmation after its critical design review in part due to adding encryption to the project following its preliminary design review. Adding encryption to the project required a change in host spacecraft to one with proper facility clearances, changes to the interface designs between LCRD and the new host, and changes to the onboard LCRD hardware to accommodate encryption.

Integration and Test

Late component deliveries have delayed integration and testing and further delays could begin to affect LCRD integration as well as spacecraft integration prior to launch. For example, payload integration and testing has been delayed by about 6 weeks due to a number of late component deliveries. In one instance, delivery of the flight support assembly—which interfaces the LCRD payload to the host spacecraft and maintains the thermal environment—was behind schedule due to late procurement of heaters and temperature sensors. In another, delivery of one of the two flight modems was late due to rework and component replacement.

Project Office Comments

LCRD project officials provided technical comments on a draft of this assessment, which were incorporated as appropriate.

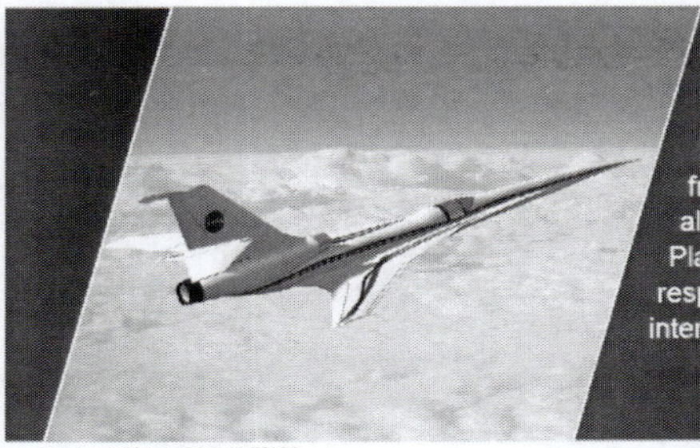

Low Boom Flight Demonstrator

LBFD is a flight demonstration project planned to demonstrate that noise from supersonic flight—sonic boom—can be reduced to acceptable levels, allowing for eventual commercial use of overland supersonic flight paths. Plans include multiple flights beyond fiscal year 2022 to gather community responses to the flights and to create a database to support development of international noise rules for supersonic flight.

Source: NASA | GAO-18-280SP

FORMULATION

- 03/13 Start of concept formulation studies
- 03/14 Mission definition review
- 09/16 LBFD formulation start
- **01/18 GAO review**
- 07/18 Preliminary design review
- 09/18 Project confirmation

IMPLEMENTATION

- 07/19 Critical design review
- 06/21 Projected first flight date
- 03/22 System acceptance review

PROJECT INFORMATION

NASA Lead Center: **None**

International Partner: **None**

Requirement Derived from: **Aeronautics Research Mission Directorate Strategic Implementation Plan**

Budget Portfolio: **Aeronautics, Integrated Aviation Systems Program**

PROJECT SUMMARY

The LBFD project entered the preliminary design and technology development phase and established preliminary cost and schedule targets in September 2016. The project expects that its preliminary cost estimate of at least $390 million will increase when it establishes its cost and schedule baseline at project confirmation planned for September 2018, due in part to the estimate not including cost and schedule reserves. The project began under a concept study effort known as the Quiet Supersonic Technology (QueSST) project and was conducted under an existing research contract. NASA selected a post-preliminary design and build contractor in March 2018. Under the QueSST effort, the project matured three of LBFD's six technologies and will need to mature its remaining three technologies before the project's preliminary design review. These include one critical technology: the design tools needed to create the aircraft's outer shape, which is necessary to meet the project's mission.

PROJECT BUDGET INFORMATION
percent of current portfolio for fiscal year 2019

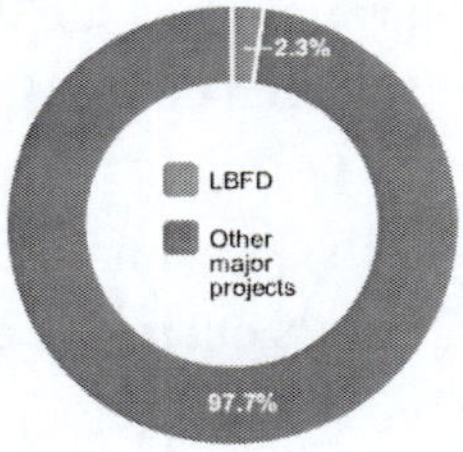

TOTAL FUNDED TO DATE
then-year dollars in millions

$75.7

PRELIMINARY COST[a]
then-year dollars in millions

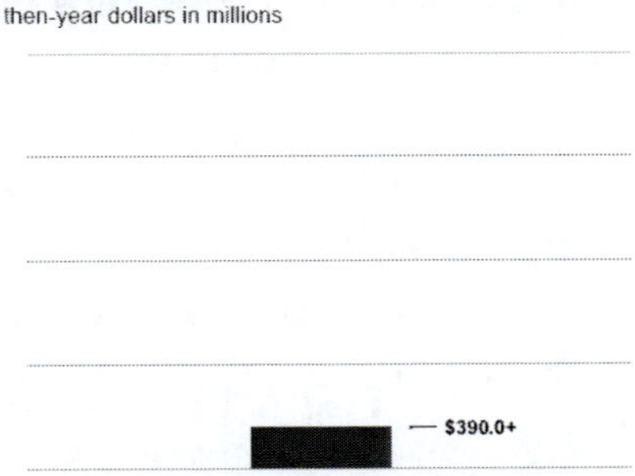

[a]This estimate is preliminary, as the project is in formulation and there is uncertainty regarding the costs associated with the design options being explored. NASA uses these estimates for planning purposes.

PRELIMINARY SCHEDULE

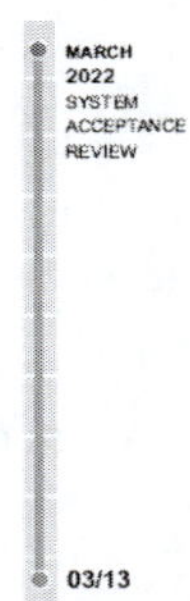

LATEST ESTIMATE
MARCH 2018

Cost and Schedule Status

The LBFD project entered the preliminary design and technology development phase and established preliminary cost and schedule targets in September 2016. The project's preliminary cost estimate is $390 million but the project expects the estimate to increase when it establishes a cost and schedule baseline at its confirmation review planned for September 2018. For example, the preliminary cost estimate does not include cost reserves.

Further, according to project officials, the cost estimate will also include an additional $39 million associated with requirements and concept study efforts—that included developing a preliminary design—conducted from 2014 to 2016. Known as the Quiet Supersonic Technology (QueSST) project, the effort was conducted under an existing research contract with Lockheed Martin. Project officials explained that QueSST used funding available each year for concept studies, but that the LBFD project is slated for dedicated funding following its preliminary design review. However, project officials do not plan to include some of the costs associated with the flights to gather community responses because a separate NASA project— the Commercial Supersonic Technology project—will lead this effort.

Additionally, the project's preliminary schedule is in flux. Since the release of the fiscal year 2019 Presidential Budget Request in February 2018, project officials indicated that dates for some key events have changed, and the system acceptance review is now scheduled 5 months later than previously planned.

Design and Technology

In March 2018, NASA selected Lockheed Martin for the post-preliminary design contract for LBFD development. Officials reported that the solicitation did not require that proposals use the preliminary design developed by Lockheed Martin under QueSST. But officials said that part of the evaluation criteria was whether the proposed concept was at a preliminary design review-level of maturity.

Based on the work completed under QueSST, the LBFD project is continuing to mature its one critical technology— the design tools used to create the aircraft's outer shape, which is necessary to achieve low-boom supersonic flight. Project officials said developing these tools and the shape, which directly impacts the sound produced by the aircraft during supersonic flight, will be the majority of the development work necessary to fly and test the LBFD aircraft. The project has assessed the design tools to be at a technology readiness level 5, which means the basic components have been integrated and tested in a simulated environment.

Three of LBFD's five heritage technologies were matured as part of the QueSST effort. The project will need to mature the remaining technologies as well as the design tools prior to its preliminary design review in July 2018. Maturing technologies during preliminary design helps reduce risks for systems entering product development. The fiber optic sensing system, which will measure bend and twist of the wings and stabilizer, and the external vision system, which includes cameras and monitors to provide forward visibility for the pilot—are not yet mature and will require additional development work. The project noted that these two technologies have flown before but not as part of a research aircraft.

Other Issues to Be Monitored

The LBFD project is using a virtual project office and is in the process of making decisions on how to execute the project using this model. The project team includes personnel from many NASA centers, allowing specialized staff in different centers to collaborate more than they would normally. The virtual project office model may highlight an organizational structure that could be beneficial for future projects, but it is too soon to tell. In addition, officials stated that the project has not decided which NASA center's policies it will follow, which include the cost and schedule reserve requirements the project will need to meet, among other guidelines.

Project Office Comments

In commenting on a draft of this assessment, LBFD project officials stated that the project is continuing with its formulation activities in fiscal year 2018. These activities will lead to authorization to proceed, with implementation following the confirmation review planned for September 2018, at which time the project will be baselined. Project officials also stated that the information contained in this project assessment reflects the most currently available projections. Project officials also provided technical comments, which were incorporated as appropriate.

Lucy

Lucy will be the first mission to investigate the Trojans, which are a population of never-explored asteroids orbiting in tandem with Jupiter. The project aims to understand the formation and evolution of planetary systems by conducting flybys of these remnants of giant planet formation. The Lucy spacecraft will first encounter a main belt asteroid—located between the orbits of Mars and Jupiter—and then will travel to the outer solar system where the spacecraft will encounter six Trojans over an 11-year mission. The mission's planned measurements include asteroid surface color and composition, interior composition, and surface geology.

Source: Southwest Research Institute (SwRI). | GAO-18-280SP

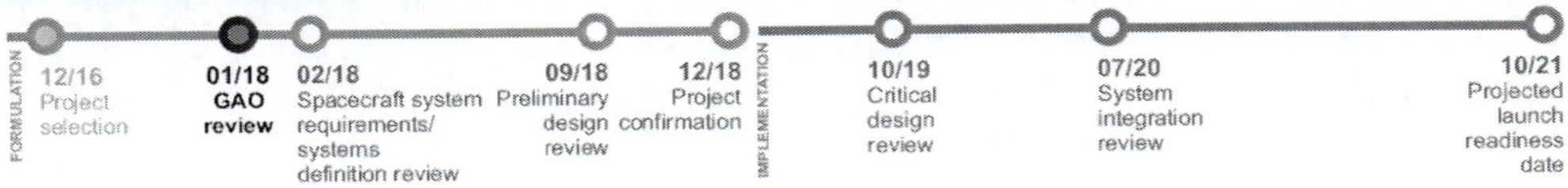

PROJECT INFORMATION

NASA Lead Center: **Goddard Space Flight Center**

International Partner: **None**

Launch Location: **Kennedy Space Center, FL**

Launch Vehicle: **TBD**

Mission Duration: **11.6 years**

Requirement Derived from: **Discovery Program Announcement of Opportunity 2014**

Budget Portfolio: **Science, Planetary Science**

PROJECT SUMMARY

In December 2016, Lucy was one of two projects selected by the Discovery program—a series of competed missions that have focused investigations and short development periods—to proceed to preliminary design and technology completion phase. Lucy is expected to have its rehearsal flyby of a main belt asteroid in 2025, and then fly by Trojan asteroids over the course of the next 8 years. Project officials stated that the design is based heavily on heritage hardware and does not require development of any critical technologies. The project is evaluating the extent to which risks are posed by existing hardware. For example, the project is tracking a risk related to the size of the solar array, which is larger than what the contractor typically builds. The project also implemented a trade on its spacecraft to address the risk of on-orbit failure from a heritage engine. Additionally, the project and NASA Launch Services have identified two launch vehicle options, but the project does not expect NASA to select a vehicle before its preliminary design review in September 2018.

PROJECT BUDGET INFORMATION
percent of current portfolio for fiscal year 2019

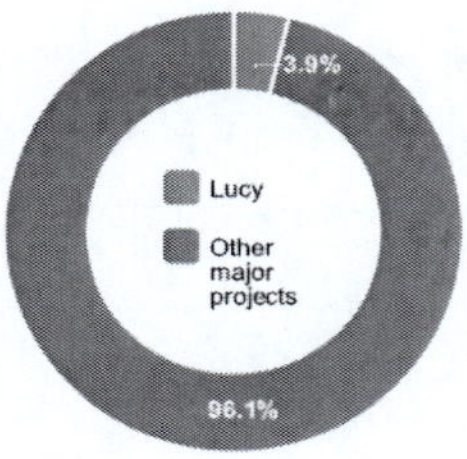

TOTAL FUNDED TO DATE
then-year dollars in millions

$57.4

PRELIMINARY COST*
then-year dollars in millions

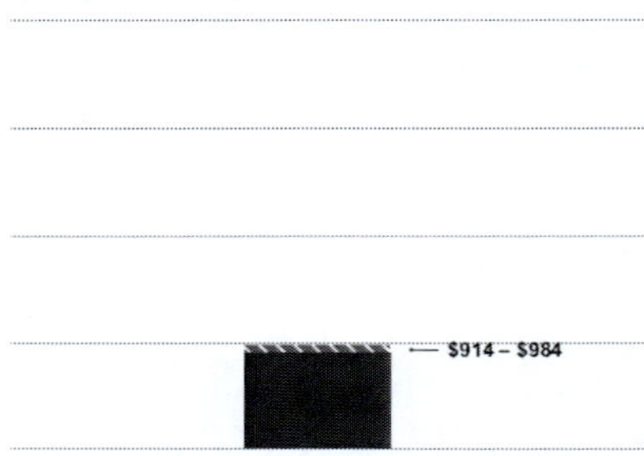

This estimate is preliminary, as the project is in formulation and there is uncertainty regarding the costs associated with the design options being explored. NASA uses these estimates for planning purposes.

PRELIMINARY SCHEDULE

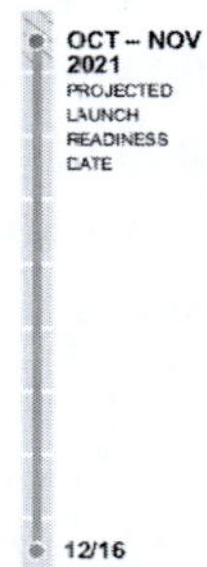

Cost and Schedule Status

In December 2016, Lucy was one of two projects selected by the Discovery program—a series of competed missions that have focused scientific investigations and short development periods—to proceed to the preliminary design and technology completion phase. At that same time, the project established preliminary cost and schedule estimates. The project's planned launch window is October- November 2021. Under that planned launch date, Lucy is expected to have its rehearsal flyby–a main belt asteroid encounter that will allow the project to test instruments—in 2025 and then fly by Trojan asteroids over the course of the next 8 years. If the project misses that launch date, another launch window opens in 2022. The project's preliminary cost range is $914 million to $984 million, and the project is currently holding cost and schedule reserves consistent with the level required by NASA center policy. The project plans to hold its preliminary design review in September 2018 and its confirmation review in December 2018, at which point it will establish its cost and schedule baseline.

Technology and Design

Project officials characterize the Lucy design as low risk because it does not require development of any critical technologies and has a high heritage design. For example, these officials stated that Lucy's design has the same architecture as prior NASA projects such as Juno and the Mars Atmosphere and Volatile Evolution Mission (MAVEN). The project plans to fly three instruments—a thermal spectrometer, a reconnaissance imager, and an imaging spectrometer—that are not exact replicas of previously flown versions but leverage heritage hardware. The Lucy project reports that its one heritage technology—an ultraflex solar array—is mature, but the project plans to use the technology in a different environment than previously demonstrated as well as make modifications that affect form, fit, and function. We have previously asserted that mature, heritage technologies must be demonstrated in a relevant environment and should be very close to form, fit, and function. The project is tracking a risk related to the size of the solar array since, at approximately 6 meters, it is larger than what the contractor typically builds. Because the contractor has not flown an array of this size, officials said they anticipate being more involved in this procurement, but there are no technical issues at this point.

According to officials, the Lucy spacecraft's proposed main engine was the same model used on a series of environmental satellites that have had engine performance problems. The engine has failed in flight more than once and was a single point failure for the Lucy project. As a result, the project completed an engine trade study in July 2017 and decided to select a different engine.

Launch

In coordination with NASA Launch Services at Kennedy Space Center, project officials identified the Atlas V and Falcon Heavy as vehicles that meet mission requirements. However, as of January 2018, the Falcon Heavy had not flown, so launch environment data was not yet available to the project. Project officials are tracking a risk that it could be necessary to rework hardware due to the Falcon Heavy launch environment. However, the project has assessed the cost impact of this scenario should the Falcon Heavy be selected. Officials said they do not expect NASA to select a launch vehicle prior to the project's planned September 2018 preliminary design review, the point at which projects prefer to select a launch vehicle. Until then, the project is designing to a mix of the two potential launch vehicles' requirements.

Project Office Comments

Lucy project officials provided technical comments on a draft of this assessment, which were incorporated as appropriate.

Mars 2020

Mars 2020 is part of the Mars Exploration Program, which seeks to further understand whether Mars was, is, or can be a habitable planet. Its rover and science instruments will explore Mars and conduct geological assessments, search for signs of ancient life, determine potential environmental habitability, and prepare soil and rock samples for potential future return to Earth. The rover will include a technology demonstration instrument designed to convert carbon dioxide into oxygen. Mars 2020 is based heavily on the Mars Science Laboratory, or Curiosity, which landed on Mars in 2012 and remains in operation.

Source: NASA/JPL-Caltech. | GAO-18-280SP

PROJECT INFORMATION

NASA Lead Center: **Jet Propulsion Laboratory**

International Partners: **Centre National d'Etudes Spatiales (France), Centro de Astrobiología and Center for the Development of Industrial Technology (Spain), Norwegian Defence Research Establishment (Norway)**

Launch Location: **Eastern Range, FL**

Launch Vehicle: **Atlas V**

Mission Duration: **2 years**

Requirement Derived from: **2011 Planetary Science Decadal Survey**

Budget Portfolio: **Science, Planetary Science**

PROJECT SUMMARY

The Mars 2020 project has encountered cost growth and schedule delays due to technical and design challenges for some components and subsystems—including a new and highly complex development—but these delays have not affected the project's overall schedule. The project did not mature all of its technologies until after the project's critical design review, which is later than recommended by best practices. Further, the project's new and complex developments continue to face technical challenges that could affect cost and schedule. Additionally, the project had an unstable design at its critical design review, with 72 percent of its drawings released. This is lower than recommended by best practices and new and highly complex developments were not stable at the time of the design review. The project is working toward system integration review in February 2018, but the project is tracking several risks that may affect schedule. For example, the new Sampling and Caching System (SCS) that will collect and cache Martian soil and rock samples is facing technical and schedule challenges.

PROJECT BUDGET INFORMATION
percent of current portfolio for fiscal year 2019

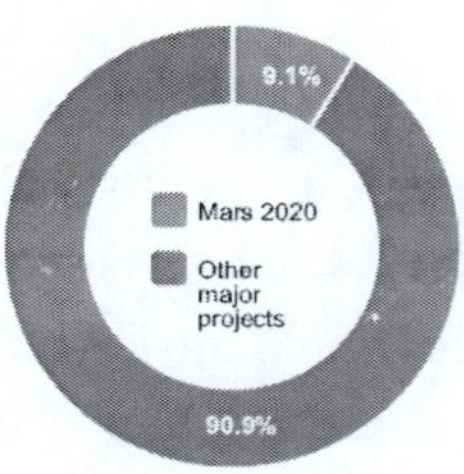

ESTIMATED FUNDING NEEDED
then-year dollars in millions

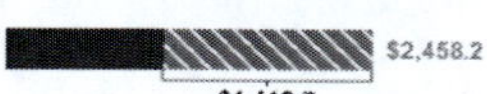

COST PERFORMANCE
then-year dollars in millions

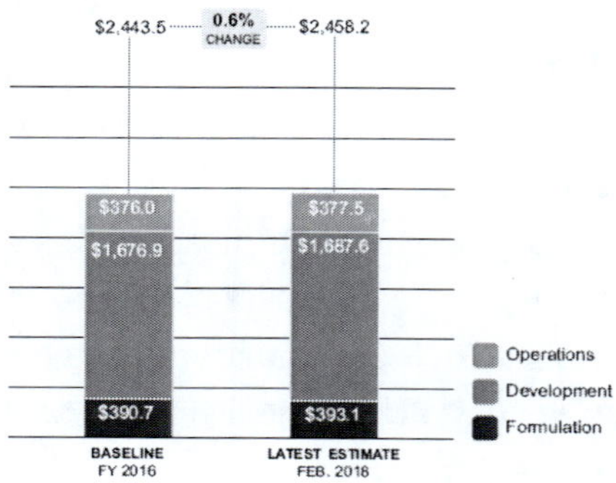

SCHEDULE PERFORMANCE

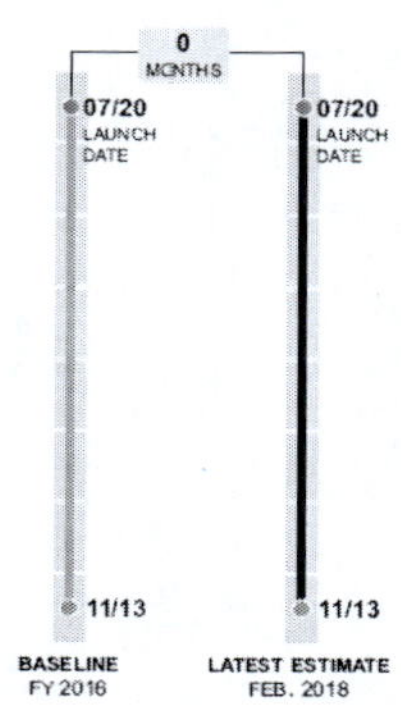

Cost and Schedule Status

The Mars 2020 project continues to meet its schedule baseline, but is not meeting the cost baseline established at its confirmation review in June 2016. The project experienced $14.7 million in cost growth due to technical challenges with a technology demonstration instrument and higher than expected integration costs for an entry, descent, and landing instrument. In addition, the project has also experienced schedule delays for some of its new developments, but these delays have not affected the project's overall schedule. For example, the project delayed its system integration review by about 3 months to February 2018 to allow for more time to build new development hardware, which were behind schedule due to technical and design challenges. Even with this delay, project officials said they are confident in the schedule established at the confirmation review because the project is holding schedule reserves at about two times the level required by Jet Propulsion Laboratory policy. In addition, the project allocated a significant portion of cost reserves to the new development items. The project is not holding cost reserves at the level required by Jet Propulsion Laboratory policy, but expects to be meeting the policy once the project receives additional funds in February 2018.

Technology

The Mars 2020 project matured its technologies later than recommended by best practices and several new and highly complex developments continue to face significant technical challenges that could affect cost and schedule. Best practices recommend maturing technologies to a technology readiness level 6 by the project's preliminary design review to help minimize risks for space systems entering product development. However, all of the technologies for the Mars 2020 project were not mature until after the program's critical design review. Further, the project is concerned late hardware deliveries and unfavorable sample contamination test results might cause the Sampling and Caching System (SCS)—the project's most challenging development that will collect and cache Martian soil and rock samples—to be delivered late to integration and test. Additionally, the project experienced technical issues with multiple components within an instrument that will search for organics and minerals that have been altered by water, which have negatively affected the instrument's cost and schedule. To mitigate these challenges, the project allocated additional reserves and brought on additional expertise. Further, the project is utilizing existing SCS development models rather than build new ones to reduce cost and schedule risk.

Design

The Mars 2020 project held its critical design review before it had a stable design, which increases the project's risk of cost growth and schedule delays during the integration and test phase. At this review, the project released about 72 percent of design drawings, which is less than the best practice of releasing 90 percent of design drawings at this review. Project officials stated that they held the design review earlier than normal to avoid delaying progress on the development of heritage technologies, which make up a large percentage of the rover. As a result of holding the review early, new and highly complex developments, such as the SCS, were not stable at the time of the design review. The unstable design contributed to delays in fabricating parts for the SCS engineering unit. As a result, the project reduced the number of engineering models it will build to help meet schedule.

The project has a series of parachute tests underway to determine whether it can fly with its heritage parachute or needs to use a strengthened parachute. These tests are to help mitigate risks related, in part, to supersonic test failures observed on an unrelated project. Officials said the project successfully tested its heritage design at supersonic speeds in October 2017, which reduces the risk that a new parachute design will be necessary. However, if future tests do not successfully demonstrate that either the heritage or redesigned parachute inflate as intended, the project may have to conduct additional development and tests, which could consume all of Mars 2020's available schedule reserve.

Other Issues to Be Monitored

The Mars 2020 project has experienced workforce shortages in several key areas, but, according to officials, the Jet Propulsion Laboratory has largely been able to meet the project's needs because it prioritized staffing for the Mars 2020 project. Other Jet Propulsion Laboratory projects have experienced workforce issues as well, which NASA attributes to the amount of work the laboratory is managing.

Project Office Comments

In commenting on a draft of this assessment, Mars 2020 project officials stated the project matured all its new technologies to the appropriate level by critical design review. Further, officials stated the project had backup technologies but none were required. Officials also stated the project has accommodated schedule delays within available schedule reserves and continues to maintain robust schedule reserve along the critical path.

Project officials also provided technical comments, which were incorporated as appropriate.

NASA ISRO – Synthetic Aperture Radar

The NASA Indian Space Research Organisation (ISRO) – Synthetic Aperture Radar (NISAR) is a joint project between NASA and ISRO that will study the solid Earth, ice masses, and ecosystems. It aims to address questions related to global environmental change, Earth's carbon cycle, and natural hazards, such as earthquakes and volcanoes. The project will include the first dual frequency synthetic aperture radar instrument, which will use advanced radar imaging to construct large-scale data sets of the Earth's movements. NISAR represents the first major aerospace science partnership between NASA and ISRO.

Source: © California Institute of Technology/Jet Propulsion Laboratory. | GAO-18-280SP

FORMULATION
- 03/14 Formulation start
- 12/14 System requirements/ mission definition review
- 06/16 Preliminary design review
- 08/16 Project confirmation

IMPLEMENTATION
- **01/18 GAO review**
- 10/18 Critical design review
- 12/19 System integration review
- 12/21 Manifested launch readiness date
- 09/22 Committed launch readiness date

PROJECT INFORMATION

NASA Lead Center: **Jet Propulsion Laboratory**

International Partner: **Indian Space Research Organisation (India)**

Launch Location: **Satish Dhawan Space Centre, India**

Launch Vehicle: **Geosynchronous Satellite Launch Vehicle Mark II**

Mission Duration: **3 years**

Requirement Derived from: **2007 Earth Science Decadal Survey**

Budget Portfolio: **Science, Earth Science**

PROJECT SUMMARY

The NISAR project continues to operate within its schedule baseline, but is not meeting its cost baseline established at its confirmation review in August 2016. The project experienced $30 million in cost growth as the result of an increase in the scope of data collection. Also, about 2 months after confirmation, the project learned that its radar reflector contractor entered into two commercial contracts, which could put a strain on the contractor's workforce and increased cost and schedule risk to the project. NASA increased near-term funding of the project in May 2017, which allowed the project to accelerate radar reflector development. In addition, the project made design changes to the radar reflector boom assembly—used to deploy the reflector when the spacecraft reaches orbit—to address risks that could prevent the boom from latching in place. The NISAR project continues to track cost and schedule risks related to its partnership with ISRO. NISAR is progressing toward its critical design review in October 2018.

PROJECT BUDGET INFORMATION
percent of current portfolio for fiscal year 2019

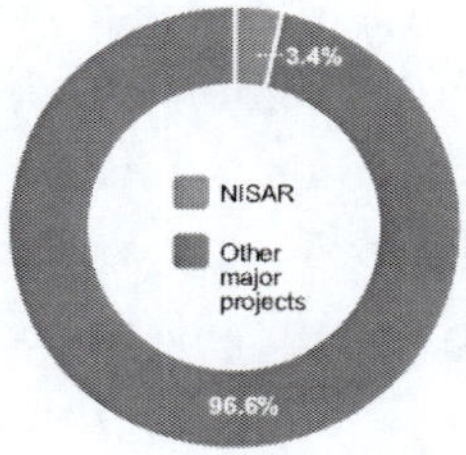

ESTIMATED FUNDING NEEDED
then-year dollars in millions

COST PERFORMANCE
then-year dollars in millions

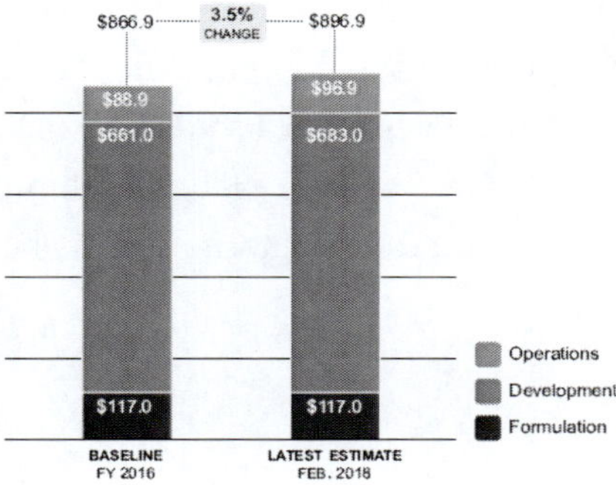

SCHEDULE PERFORMANCE

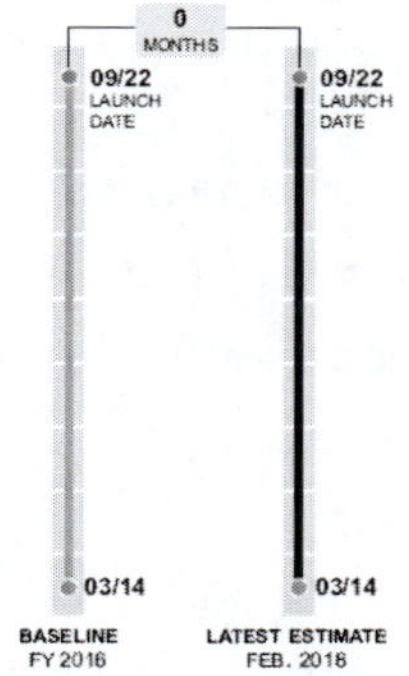

Cost and Schedule Status

The NISAR project continues to operate within its schedule baseline, but is not meeting the cost baseline established at its confirmation review in August 2016. The project experienced $30 million in cost growth as the result of an increase in the scope of data collection in response to additional data needs being identified by an interagency working group. The additional data include soil moisture and natural hazard data that would be of value for other federal agencies and the science community. Also, about 2 months after confirmation, the project learned that its radar reflector contractor entered into two commercial contracts, which could put a strain on the contractor's workforce and increased cost and schedule risk to the project. To avoid schedule conflicts with the contractor's new contracts, NASA increased the project's near-term funding in May 2017 so that the project could accelerate development of the radar reflector. NISAR is progressing toward its critical design review in October 2018 and is currently holding cost and schedule reserves consistent with the level required by Jet Propulsion Laboratory policy.

Design

The NISAR project made design changes to the radar reflector boom assembly—used to deploy the radar reflector when the spacecraft reaches orbit—to address risks that could prevent the boom from latching in place. Specifically, the project modified how the boom is deployed, including using spring dampers to deploy the boom instead of a cable-driven system and non-explosive launch restraints to reduce shock levels. Project officials said the boom redesign has single point failures, but the probability of failure is lower compared to the previous design.

The NISAR project continues to track the system's mass as a significant risk. Maintaining adequate mass margins is a key indicator of design stability. Project officials stated that they continue to improve their knowledge of mass through the use of engineering models and more matured design, and sometimes this has led to an increase in mass. For example, the project used computer aided design to determine the length of cabling needed and found this increased its mass estimate over prior estimates. In July 2017, the project instituted a mass control plan, which project officials said would require all increases in mass to be reported and determinations be made regarding the necessity of the change.

Developmental Partner

The NISAR project continues to track a risk that process differences between NASA and its development partner, the Indian Space Research Organisation (ISRO), could negatively affect cost and schedule, but a recent project assessment concluded that collaboration between the two organizations has been effective. For example, in July 2017, the project signed an updated cooperative project plan that outlines how the two organizations should interface on topics such as requirements and technical information.

The project is also tracking a risk related to the ISRO- provided launch vehicle—the Geosynchronous Satellite Launch Vehicle (GSLV) Mark II—which must meet all NASA-ISRO agreed-upon criteria before it can be used.

For example, the GSLV Mark II must have three successful launches and one successful 4-meter fairing launch prior to NISAR's planned launch date in 2021. As of January 2018, ISRO had four consecutive successful GSLV Mark II launches, but had not yet planned a 4-meter fairing launch.

Project Office Comments

NISAR project officials provided technical comments on a draft of this assessment, which were incorporated as appropriate.

Orion Multi-Purpose Crew Vehicle

The Orion Multi-Purpose Crew Vehicle (Orion) is being developed to transport and support astronauts beyond low-Earth orbit, including traveling to Mars or an asteroid. The Orion program is continuing to advance development of the human safety features, designs, and systems started under the Constellation program, which was canceled in 2010. Orion is planned to launch atop NASA's Space Launch System (SLS). The current design of Orion consists of a crew module, service module, and launch abort system.

Source: © 2017 Lockheed Martin Corporation | GAO-18-280SP

- 07/06 Formulation start under Constellation program
- 02/10 Orion canceled under Constellation program
- FORMULATION
- 02/12 Formulation start
- 08/14 Preliminary design review
- 12/14 Exploration flight test
- 09/15 Project confirmation
- IMPLEMENTATION
- 10/15 Critical design review
- **01/18 GAO review**
- 12/19-06/20 First uncrewed flight replanned launch readiness date
- 04/23 First crewed flight committed launch readiness date

PROJECT INFORMATION

NASA Lead Center: **Johnson Space Center**

International Partner: **European Space Agency**

Launch Location: **Kennedy Space Center, FL**

Launch Vehicle: **Space Launch System**

Mission Duration: **Up to 21 day active mission duration capability with four crew**

Requirement Derived from: **NASA Authorization Act of 2010**

Budget Portfolio: **Exploration, Exploration Systems Development**

PROJECT SUMMARY

The Orion program continues to operate within its schedule baseline but NASA expects the program to exceed its cost baseline through Exploration Mission-2 (EM-2) due to new hardware and the program addressing development challenges. The extent of cost growth is unknown, but NASA plans to complete a new cost estimate by June 2018. For example, according to NASA, the cost increases have been driven in part by moving from a single-piece, or monolithic, heatshield design to one that employs blocks in order to improve its structural strength. The program's service module—contributed by the European Space Agency—is currently driving the program schedule as well as the launch schedule for the first mission. The module has proven more difficult to produce than expected and has been delayed numerous times. The service module is at risk of further delays that may affect NASA's planned launch date for EM-1, which would begin to consume schedule reserve for EM-2. In addition, the program is addressing some parts material failures. Specifically, the spacecraft's avionics cards did not withstand vibration testing and required a materials change.

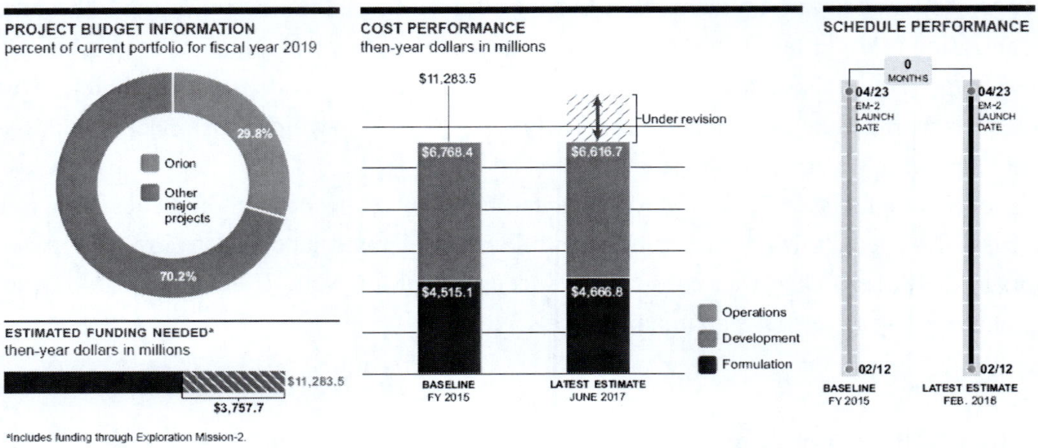

Cost and Schedule Status

The Orion program continues to operate within its schedule baseline but NASA expects the program to exceed its cost baseline through EM-2, and is at risk for future schedule delays. The project's life-cycle cost estimate is expected to increase beyond its $11.28 billion baseline—due in part to the EM-1 schedule delay—when a planned, new cost estimate for the program is complete in June 2018. According to preliminary analysis, major drivers of the potential cost increase also include new hardware and addressing development challenges. For example, there was a cost impact when the program moved from a single-piece, or monolithic, heatshield design to one that employs blocks in order to improve its structural strength.

In December 2017, NASA announced December 2019 as the new internal launch readiness date for EM-1 and that the agency also allocated 6 months of schedule reserve to June 2020 for possible manufacturing and production schedule risks. While the Orion program did not have a committed launch date for EM-1, the recent delay has reduced the amount of time available to the program between EM-1 and its committed EM-2 launch date of April 2023. In addition, the delay means that the program will continue to consume resources for EM-1 that would have otherwise been available for development on EM-2, thus increasing pressure on the EM-2 cost and schedule. If NASA delays EM-1 beyond December 2019—which is likely given both Orion and the Space Launch System have no schedule margin to meet their deliveries for this date—the schedule reserve for Orion's committed EM-2 launch date of April 2023 would continue to erode and put the program at risk for future schedule delays.

Developmental Partner

The late completion and delivery of the European Service Module (ESM)—a European Space Agency contribution via agreement with NASA—is driving the program's schedule and may further delay EM-1. The European Space Agency has delayed delivery of the service module 14 months since the element's critical design review in June 2016. The ESM currently has no schedule reserve to support the December 2019 launch schedule, meaning that any additional delays will compress or delay integration activities prior to launch. Further, NASA is tracking a risk that the ESM could be delayed beyond the current estimated delivery date of June 2018. Such a delay would likely delay the EM-1 launch date beyond December 2019.

Program officials stated that recent ESM delays are due in part to late component deliveries from subcontractors, especially valves. However, they also noted at least one of those valves is currently not meeting specifications, indicating that the design was not sufficiently mature prior to production.

The valve, a heritage design from the Space Shuttle program that maintains pressure in the propulsion system, is not sealing properly due to the increased pressures necessary on the ESM. The program has previously stated that all sides believed that the development of the ESM would be easier than it has proven to be, being based on a prior European Space Agency spacecraft. However, the changes have been more substantial than expected and the production of the first flight unit has faced setbacks. The program stated that they expect the production of the flight unit for the second flight to be quicker, though it will require some additional elements to support crew.

Technology and Design

Avionics design has recently become an issue for the Orion program and is currently 2 months behind the ESM in terms of driving the program's schedule. The avionics cards, along with many other onboard systems, have to withstand the vibration and radiation environment on Orion. The program found that the circuits on the avionics cards were cracking under operating conditions due to a poor design. As a result, officials stated that the program determined the root cause of the failure and redesigned the cards' base material to better withstand the environment. The program worked to reorganize the integration and test schedule to allow for the replacement of the cards.

Program Office Comments

In commenting on a draft of this assessment, Orion program officials stated that they believe the risk- informed approach that they use to address and resolve issues has proven to be successful. Also, program officials stated the program remains on track to meet its

April 2023 baseline for EM-2. Program officials also provided technical comments, which were incorporated as appropriate.

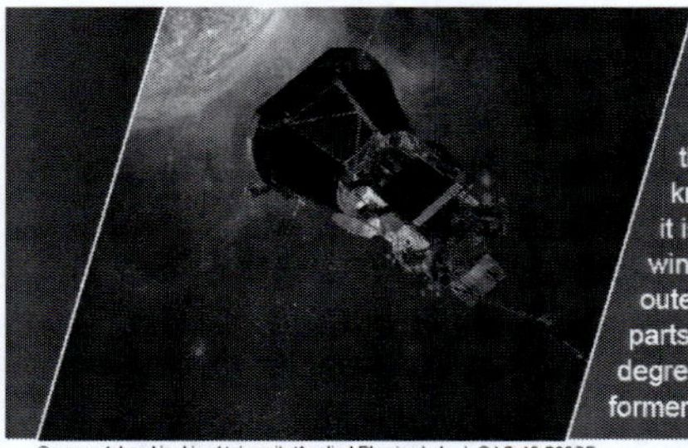

Parker Solar Probe

PSP will be the first NASA mission to visit a star. Using the gravity of Venus, the spacecraft will orbit the Sun 24 times and gather information to increase knowledge about the solar wind, including its origin, acceleration, and how it is heated. PSP instruments will observe the generation and flow of solar winds from very close range and sample and take measurements of the Sun's outer atmosphere, where solar particles are energized. To achieve its mission, parts of the spacecraft must be able to withstand temperatures exceeding 2,500 degrees Fahrenheit and endure blasts of extreme radiation. The project was formerly named Solar Probe Plus, or SPP, and was renamed in May 2017.

Source: Johns Hopkins University/Applied Physics Lab. | GAO-18-280SP

FORMULATION: 12/09 Formulation start; 11/11 System requirements/ mission definition review; 01/14 Preliminary design review; 03/14 Project confirmation

IMPLEMENTATION: 03/15 Critical design review; 05/16 System integration review; **01/18 GAO review**; 08/18 Committed launch readiness date

PROJECT INFORMATION

NASA Lead Center: **Goddard Space Flight Center**

International Partner: **None**

Launch Location: **Cape Canaveral Air Force Station, FL**

Launch Vehicle: **Delta IV-heavy class with NASA-provided upper stage**

Mission Duration: **7 years**

Requirement Derived from: **2012 Heliophysics Decadal Survey**

Budget Portfolio: **Science, Heliophysics**

PROJECT SUMMARY

The PSP project continues to target an August 2018 planetary launch, but the project is encountering a number of technical issues that may lead to a schedule delay or result in the project launching with fewer capabilities than originally planned. The project will need to resolve the following three issues by spring 2018 to maintain the launch window. First, the Solar Probe Cup, which is one component of an instrument package necessary to meet top-level mission requirements, has encountered several technical issues during development. Some of these issues may require the project to descope this instrument. Second, the project is completing qualification testing of the separation system after several separation bolts in the launch vehicle interface failed during a test. Lastly, the project has identified contamination stemming from an alloy used in three locations on two instruments. In one of the locations, the project must determine if it can limit the contamination. If it cannot, the project will need to delay its launch in order to replace the affected part.

PROJECT BUDGET INFORMATION
percent of current portfolio for fiscal year 2019

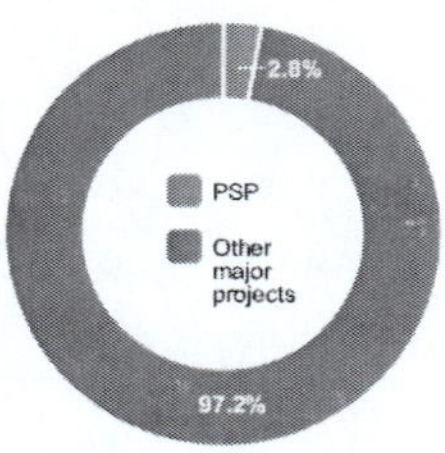

ESTIMATED FUNDING NEEDED
then-year dollars in millions

COST PERFORMANCE
then-year dollars in millions

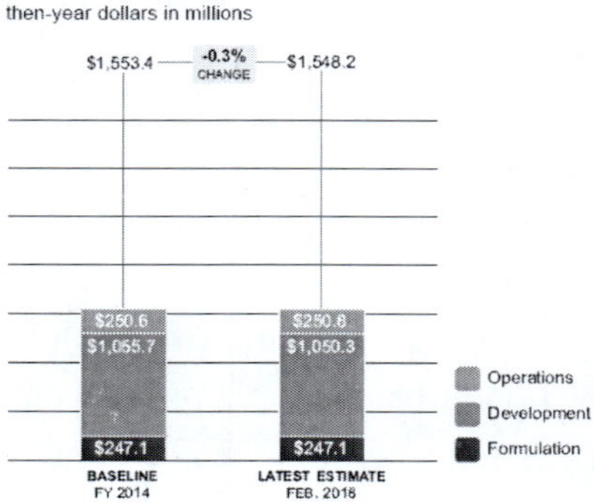

SCHEDULE PERFORMANCE

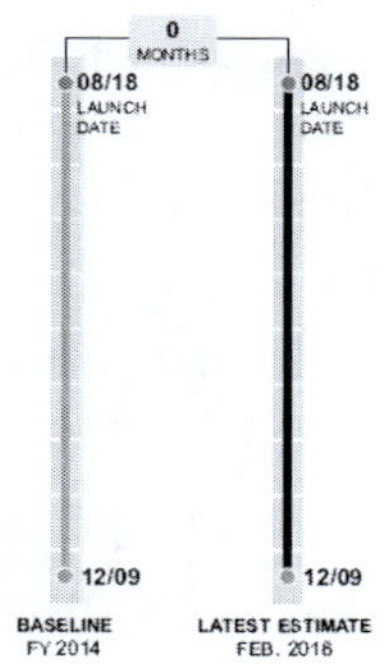

Cost and Schedule Status

The PSP project continues to target an August 2018 planetary launch, but the project is encountering a number of technical issues that may lead to a schedule delay. Maintaining the project's 2018 launch window is important because a potential window only opens every 10 months. The 2019 launch window would result in a longer mission duration and require more fuel, and after that, the next window that meets requirements is 2023.

The project continues to hold schedule reserves at Applied Physics Laboratory-recommended levels, but the project is tracking a risk that there may not be adequate reserves to address any future issues that may arise. The project is also tracking a risk that it may exhaust its cost reserves in fiscal year 2018 addressing instrument issues and retaining project staff, which could lead to the need for additional headquarters-held cost reserves.

Launch

In September 2017, while testing the interface between the launch vehicle and the spacecraft, three of the six separation nuts failed to release their bolts. If this occurred during launch, it would result in a total mission failure. NASA's Launch Services Program, which obtained launch services for PSP, initiated an anomaly investigation. This investigation determined that the bolts were improperly installed. The investigation board identified corrective actions, which have passed initial tests. The separation system plan includes completing qualification testing by April 2018 and includes schedule margin. However, if additional issues are identified, the project could potentially miss the 2018 launch window.

Integration and Test

The Solar Probe Cup (SPC), which is part of an instrument package necessary to meet top-level mission requirements to gather information about particles in the solar wind, has encountered several technical issues during integration and testing. For example, recent testing has identified scenarios where the spacecraft's different operating temperature environments could result in twisting between the SPC and spacecraft, which could lead to cracks over time. To mitigate this risk, the project is conducting testing to determine the scope of this issue. If twisting could occur repeatedly throughout the mission, the project will consider de-scoping the SPC, which would require approval from NASA. The project plans to make a decision in March 2018 about whether to fly the SPC.

Other Issues to Be Monitored

The project is tracking a risk that an alloy used in several locations on the spacecraft will release gases when exposed to the high temperatures found where the spacecraft is intended to operate. The released gases can later re-solidify and contaminate the spacecraft. The alloy is found in three locations, supporting two instrument suites, on the spacecraft—the four FIELDS whip antennas, their respective thermal shields, and the SPC thermal shield— which are required to meet top-level mission requirements. Testing to understand the alloy's performance revealed that the alloy released gases even at temperatures much cooler than where the spacecraft will operate. The project is pursuing two mitigations. First, it is conducting tests to develop a contamination model, which should indicate the effects, if any, the re-solidified gases have on the spacecraft and help project officials determine if it is safe to fly the spacecraft with the existing alloy. Second, officials told us that they have ordered new material which could be used to replace the four FIELDS whip antennas and the SPC thermal shield. The project plans to make a decision if they will replace existing parts with the new material by February 2018. The project cannot replace the FIELDS thermal shields, so they have designed and implemented an additional shield, which will undergo testing in March 2018.

Project Office Comments

PSP project officials provided technical comments on a draft of this assessment, which were incorporated as appropriate.

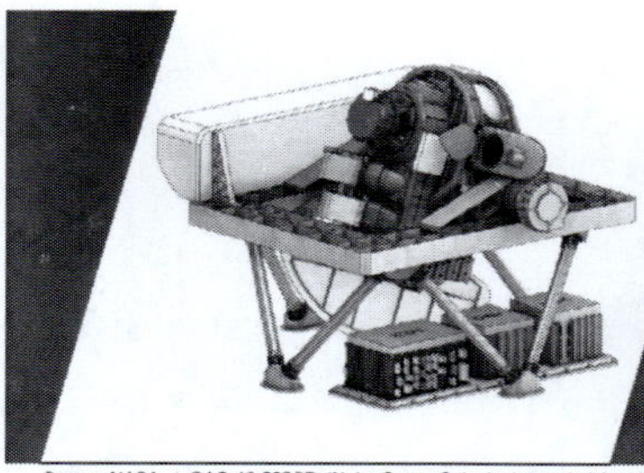

Plankton, Aerosol, Cloud, ocean Ecosystem

The Plankton, Aerosol, Cloud, ocean Ecosystem (PACE) is a polar-orbiting mission that will use advanced global remote sensing instruments to improve scientists' understanding of ocean biology, biogeochemistry, ecology, aerosols, and cloud properties. PACE will extend climate-related observations begun under earlier NASA missions, which will enable researchers to study long-term trends on Earth's oceans and atmosphere, and ocean-atmosphere interactions. PACE will also enable assessments of air and coastal water quality, such as the locations of harmful algae blooms.

Source: NASA | GAO-18-280SP (Note: Ocean Color Instrument pictured above.)

FORMULATION
- 06/16 Formulation start
- 01/17 System requirements/ mission definition review
- **01/18 GAO review**
- 03/19 Preliminary design review
- 06/19 Project confirmation

IMPLEMENTATION
- 01/20 Critical design review
- 05/21 System integration review
- 08/22 Projected launch readiness date

PROJECT INFORMATION

NASA Lead Center: **Goddard Space Flight Center**

International Partner: **None**

Launch Location: **TBD**

Launch Vehicle: **TBD**

Mission Duration: **3 years**

Requirement Derived from: **2007 Earth Science Decadal Survey**

Budget Portfolio: **Science, Earth Science**

PROJECT SUMMARY

The PACE project entered the preliminary design and technology completion phase in July 2017, but is facing funding uncertainty. The fiscal year 2018 President's Budget Request did not include funding for PACE. In order to meet scientific requirements and stay within NASA's established $805 million cost cap, the PACE project modified capabilities and made system trades. The project used a design-to-cost process, which requires the project to determine what baseline capabilities are achievable within cost limitations. The PACE mission cost cap continues to create risk around the project's launch vehicle selection, but officials are mitigating this risk by exploring alternative launch options, such as a ride share agreement with the Air Force. Project officials said they expect NASA to select a launch vehicle before project confirmation, currently planned for June 2019.

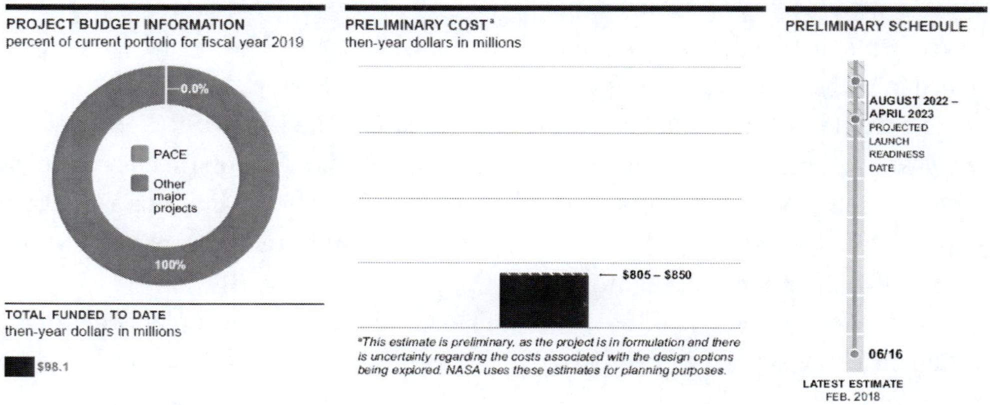

Cost and Schedule Status

In July 2017, NASA approved the PACE project to enter the preliminary design and technology completion phase, but with funding uncertainty. NASA did not request funding for PACE in its fiscal year 2018 budget request— an action attributed to budget constraints and higher science mission priorities. NASA also did not request funding for PACE in its fiscal year 2019 budget request. Despite proposed cancellation, the project is working toward preliminary design review scheduled for March 2019 and is continuing to use the design-to-cost process that requires the project to determine a baseline set of capabilities that are achievable within the project's $805 million mission cost cap at the 65 percent confidence level. This includes the $705 million allocated to the project and $100 million allocated to NASA headquarters for science-related activities, such as the calibration and validation of instrument data and processing of science data. The project is holding cost and schedule reserves consistent with NASA center policy.

Technology and Design

The PACE project modified capabilities and made system trades to meet scientific requirements within NASA's established $805 million cost cap. As part of the design-to-cost process, the project added capability to its primary instrument—the Ocean Color Instrument (OCI)—including extending performance to a new ultraviolet range and adding components for daily calibration. Additionally, the project originally sought an agreement with the Indian Space Research Organisation (ISRO) for a contributed polarimeter because the cost of procuring the instrument was higher than the amount allocated under the project's cost cap. However, according to officials, ISRO's inability to meet schedule needs and the uncertainty surrounding the future of the project led ISRO to withdraw its concept of a contributed polarimeter. The project is now pursuing two smaller contributed instruments that have mature designs from the University of Maryland-Baltimore County and the Netherlands Space Office—the Dutch space agency. According to officials, the project would use these smaller instruments in place of the larger ISRO polarimeter. While the ISRO polarimeter was a more capable instrument, the two polarimeters are expected to provide substantive scientific information for the cloud and atmospheric science community. The project will continue to make capability trades as part of the design-to-cost approach until project confirmation.

The PACE project made a design decision to tilt the OCI on a platform as opposed to tilting the entire spacecraft in order to meet a requirement that the OCI must tilt during ome points of operation to avoid sun glint, or reflection of the sun off the ocean that causes a loss in data. The project came to this decision through a trade study, which, according to officials, demonstrated that tilting the entire spacecraft would take longer than tilting just the platform, lessening the amount of time the project has to collect data. However, the project is now tracking a risk that the inertia created by the OCI platform could interfere with data collection, which the project plans to mitigate by optimizing the placement and alignment of the OCI on its platform.

Launch

The PACE project is pursuing a shared ride agreement with the Air Force, which could help to mitigate a launch vehicle risk that the project is tracking. The launch vehicle cost remains the project's top risk, which could cause the project to exceed the $705 million allocated to the project or have to reduce its science capabilities. According to officials, a shared ride option would allow PACE to be the primary mission with a secondary Air Force payload at launch, which would reduce the project's launch vehicle costs and free up funding for other areas. The project has a launch vehicle budget of $105 million, and under a potential shared ride agreement, the project would be required to cover only a portion of the total vehicle costs. Project officials said they expect NASA to select a launch vehicle before project confirmation, currently planned for June 2019.

Project Office Comments

PACE project officials provided technical comments on a draft of this assessment, which were incorporated as appropriate.

Psyche

Psyche will be the first mission to visit a metal asteroid and aims to understand a previously unexplored component of the early building blocks of planets: iron cores. The project plans to orbit the Psyche asteroid to determine if it is a planetary core, characterize its topography, assess the elemental composition and determine the relative ages of its surface regions.

Source: NASA/JPL-Caltech/Arizona State Univ./Space Systems Loral/Peter Rubin. | GAO-18-280SP

FORMULATION
12/16 Project selection
01/18 GAO review
04/18 Project mission system review
03/19 Preliminary design review
05/19 Project confirmation
IMPLEMENTATION
04/20 Critical design review
12/20 System integration review
08/22 Projected launch readiness date

PROJECT INFORMATION

NASA Lead Center: **Jet Propulsion Laboratory**

International Partner: **None**

Launch Location: **Cape Canaveral Air Force Station, FL**

Launch Vehicle: **TBD**

Mission Duration: **21 months science operation**

Requirement Derived from: **Discovery Program Announcement of Opportunity 2014**

Budget Portfolio: **Science, Planetary Science**

PROJECT SUMMARY

In December 2016, Psyche was one of two projects selected by the Discovery program—a series of competed missions that have focused scientific investigations and short development periods—to proceed to the preliminary design and technology completion phase. NASA selected the project's 2023 launch proposal, but later directed the project to work to an accelerated launch readiness date of August 2022. The accelerated launch date will allow Psyche to arrive at the asteroid over 4 years earlier than the original timeline due to a quicker flight. According to project officials, the Psyche project's current design utilizes mature, heritage technologies with some modifications. The project plans to fly three instruments that have flown on prior planetary missions and buy a commercially available spacecraft design. The project also plans to fly the Deep Space Optical Communications technology demonstration (DSOC), which is a laser-based communication device that could be beneficial to future deep space missions requiring high data rates, but it is not needed to meet Psyche's science requirements. As a result, Psyche could launch without DSOC if it experiences delays.

PROJECT BUDGET INFORMATION
percent of current portfolio for fiscal year 2019

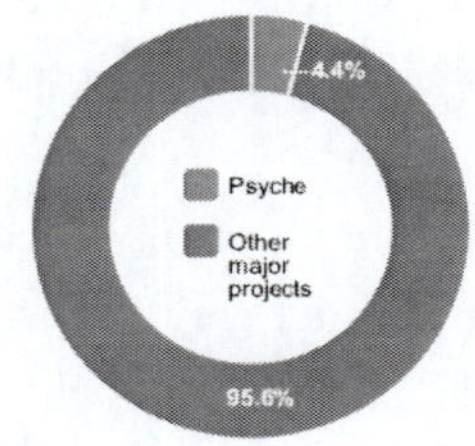

TOTAL FUNDED TO DATE
then-year dollars in millions

$50.3

PRELIMINARY COST[a]
then-year dollars in millions

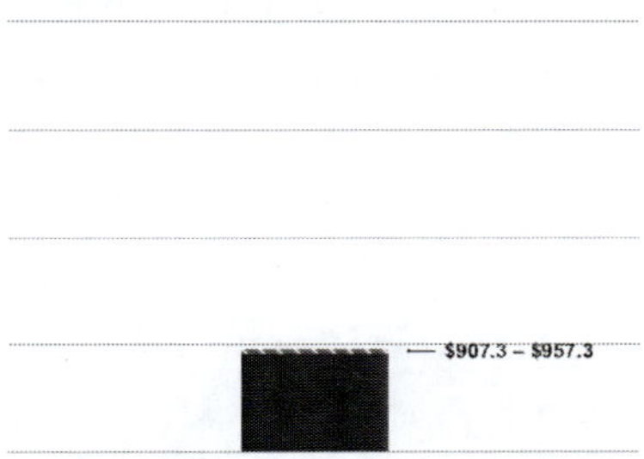

[a]*This estimate is preliminary, as the project is in formulation and there is uncertainty regarding the costs associated with the design options being explored. NASA uses these estimates for planning purposes.*

PRELIMINARY SCHEDULE

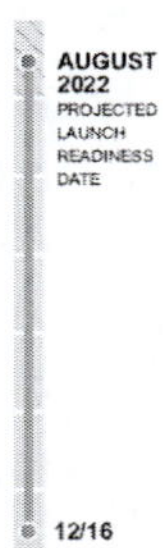

LATEST ESTIMATE
FEB. 2018

Cost and Schedule Status

In December 2016, Psyche was one of two projects selected by the Discovery program—a series of competed missions that have focused scientific investigations and short development periods—to proceed to the preliminary design and technology completion phase. NASA selected the project's 2023 launch proposal, but later directed the project to work to an accelerated launch readiness date of August 2022. The accelerated launch will allow Psyche to arrive at the asteroid over 4 years earlier than the original timeline due to a quicker flight path. At the most recent decision point, the project set a preliminary cost range of $907.3 million to $957.3 million. The project plans to hold its preliminary design review in March 2019 and its confirmation review in May 2019, at which point it will formally establish its cost and schedule baseline. The project is currently holding cost and schedule reserves consistent with the level required by Jet Propulsion Laboratory policy.

Technology and Design

The Psyche project reported that its current design does not use any critical technologies and is based heavily on heritage technologies with modifications, which project officials assess as all being matured to at least technology readiness level 6. We have previously asserted that mature technologies must be demonstrated in a relevant environment and should be very close to form, fit, and function. The Psyche project reports that its heritage technologies are mature, but the project plans to modify some technologies. For example, the project plans to modify the Gamma Ray Neutron Spectrometer instrument—previously used on a mission to Mercury and will be used to determine Psyche's elemental composition—by adding a new cooling system, which the project is tracking as a risk because of possible problems accommodating the new cooling system. Project officials stated the spacecraft bus design—reported mature by the project—is based on a commercially available design that is used for Earth-orbiting communication satellites. The project plans to make minor modifications to some of the spacecraft bus to enable it to operate more robustly in deep space.

The project also plans to fly the Deep Space Optical Communications technology demonstration (DSOC), which is a laser-based communication device that could be beneficial to future deep space missions requiring high data rates. NASA is developing and funding DSOC as a separate project in the Space Technology Mission Directorate. As a result, the Psyche project does not control the cost or schedule for DSOC. The Psyche and DSOC projects have been working closely together to align their schedules, such as by establishing regular team meetings and developing a memorandum of understanding to ensure consistent expectations between both projects. If DSOC experiences delays, project officials stated that there is an option that Psyche could launch without it because DSOC is not needed to meet Psyche's science requirements.

Other Issues to Be Monitored

In September 2017, the project reported a risk that it may have to conduct integration and testing off-site because it is planning to share a clean room with the Europa Clipper project, which has stricter planetary protection and contamination control requirements. These stricter requirements have a cost impact and the project is researching options to partition the clean room without jeopardizing the Europa Clipper project's requirements. The project is also researching options to conduct integration and testing off-site, such as using the contractor's facilities.

Project Office Comments

Psyche project officials provided technical comments on a draft of this assessment, which were incorporated as appropriate.

Radiation Budget Instrument

The Radiation Budget Instrument (RBI) is a scanning radiometer that NASA planned to launch on the National Oceanic and Atmospheric Administration's (NOAA) Joint Polar Satellite System 2 (JPSS-2). RBI's planned mission was to support global climate monitoring by continuing measurements of the Earth's reflected sunlight and emitted thermal radiation made by NASA and NOAA satellites over the past 30 years. This data was intended to represent one of two key sets of measurements needed to determine whether the Earth is warming or cooling.

Source: Harris Corporation. | GAO-18-280SP

FORMULATION
- 07/14 Formulation start
- 12/14 System requirements review
- 05/16 Preliminary design review
- 07/16 Project confirmation

IMPLEMENTATION
- 09/17 Critical design review
- **01/18 GAO review**
- 01/18 Project cancellation

PROJECT INFORMATION

NASA Lead Center: **Langley Research Center**

International Partner: **None**

Launch Location: **Not applicable**

Launch Vehicle: **Not applicable; instrument hosted on JPSS-2 spacecraft**

Mission Duration: **7 years**

Requirement Derived from: **NASA Strategic Plan**

Budget Portfolio: **Science, Earth Science**

PROJECT SUMMARY

According to NASA's cancellation memorandum, NASA canceled the RBI project because of continued cost growth, technical issues, and poor contractor performance. The RBI project continued to experience contract cost growth, and as of November 2017, contract costs for design, fabrication, and delivery of RBI had grown by about 110 percent from the original value for this work, or an increase of $115 million. This increase depleted all remaining project-held cost reserves as the project entered the riskiest point of the development cycle. Ultimately, NASA determined that continuing to fund RBI from within the Earth Science Division budget would slow other important activities.

PROJECT BUDGET INFORMATION
percent of current portfolio for fiscal year 2019

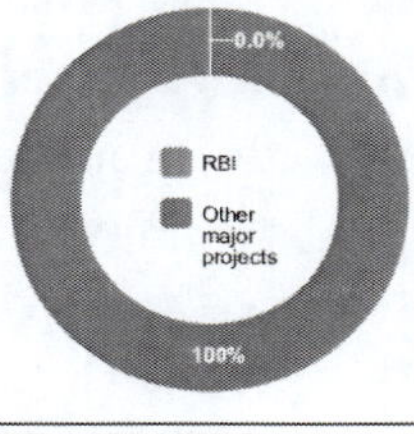

ESTIMATED FUNDING NEEDED
then-year dollars in millions

COST PERFORMANCE
then-year dollars in millions

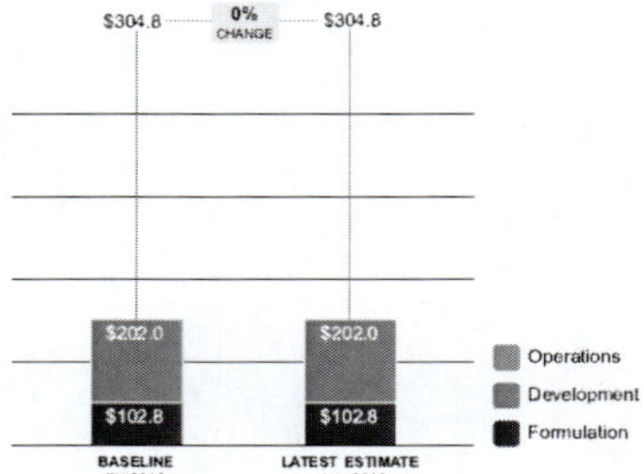

SCHEDULE PERFORMANCE

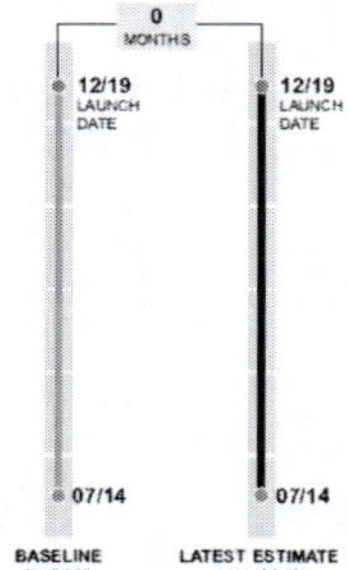

Project Status

According to NASA's cancellation memorandum, the RBI project was canceled because of continued cost growth, technical issues, and poor contractor performance.

The project identified the following specific issues that contributed to poor project cost and schedule performance: schedule planning that did not realistically account for contractor past performance, insufficient oversight of the contractor and its subcontractors, and poor integration of government and contractor teams with limited transparency concerning risks and issues. In 2017, both the government and the contractor took steps to respond to poor project performance, including changes to both the government and contractor program management teams, but ultimately NASA determined that continuing to fund RBI from within the Earth Science Division budget would slow other important activities. Based on the historical performance of other instruments and assessments of the likely remaining lifetime of on-orbit satellites, NASA believes the risk of a data gap is low if the next radiation budget-measuring instrument launches in 2027.

Cost and Schedule Status

The RBI project continued to experience contract cost growth, and was likely to exceed its cost baseline. As of November 2017, contract costs for design, fabrication, and delivery of RBI had grown by about 110 percent from the original value for this work, or an increase of $115 million. This cost increase depleted all remaining project-held cost reserves as the project entered the riskiest point of the development cycle—the integration and test phase. In 2016 and 2017, we found that projects appear most likely to rebaseline during this period.[47] In November 2017, the project requested additional reserves to cover increased contractor costs. If NASA would have approved the increase, it would have brought the project's costs to within approximately $2 million of its established cost baseline, which officials said was likely to be exceeded.

To support the decision of whether or not to continue the project, the project updated its joint cost and schedule confidence level, which is the likelihood that a project will meet its cost and schedule baseline. The update indicated that the project had a 30 percent chance of meeting the original delivery date of April 2019. As a result, the project updated the delivery date with NOAA to July 2019. The confidence level update showed that RBI had

---

[47] GAO-16-309SP and GAO-17-303SP.

a 70 percent chance of meeting the later delivery date, but it would likely require additional funding in excess of the project's cost baseline if other technical issues arose.

Technology and Design

The RBI project held its critical design review in September 2017 with 100 percent of its design drawings released. Best practices show that releasing 90 percent of design drawings by critical design review lowers the risk of subsequent cost and schedule growth. The review, previously scheduled for June 2017, was delayed to provide more time to resolve technical issues and to have additional test results available to inform the review board on the project's readiness to proceed. More specifically, the project conducted testing on an engineering development unit as a risk reduction measure and identified technical issues with the telescope and various electronic components that needed to be addressed prior to the review.

Project Office Comments

In commenting on a draft of this assessment, RBI project officials stated that RBI was not included in the President's fiscal year 2018 budget request and continuing to fund RBI from within the Earth Science Division budget would slow other important activities. Project officials stated that multiple efforts by the RBI government and contractor team to improve performance were yielding positive results, including the resolution of technical issues leading to a successful critical design review. Officials also said improved cost and schedule controls were also proving to be effective. Project officials also provided technical comments, which were incorporated as appropriate.

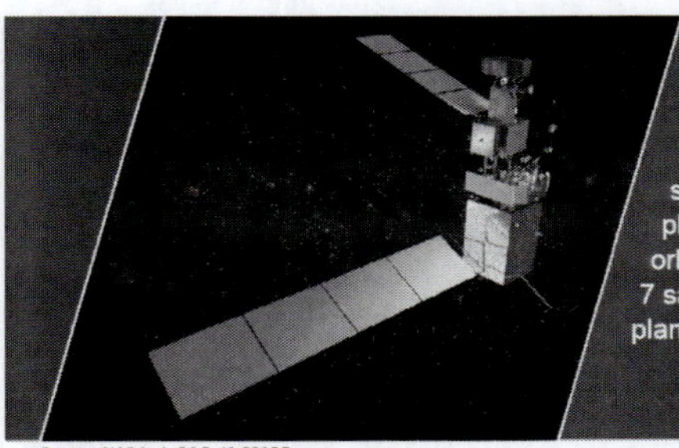

Restore-L

The Restore-L project will demonstrate the capability to refuel on-orbit satellites for eventual use by commercial entities. Specifically, Restore-L plans to autonomously rendezvous with, inspect, capture, refuel, adjust the orbit of, safely release, and depart from the U.S. Geological Survey's Landsat 7 satellite. Landsat 7 can extend operations if successfully refueled, but it is planned for retirement if the technology demonstration is unsuccessful.

Source: NASA | GAO-18-280SP

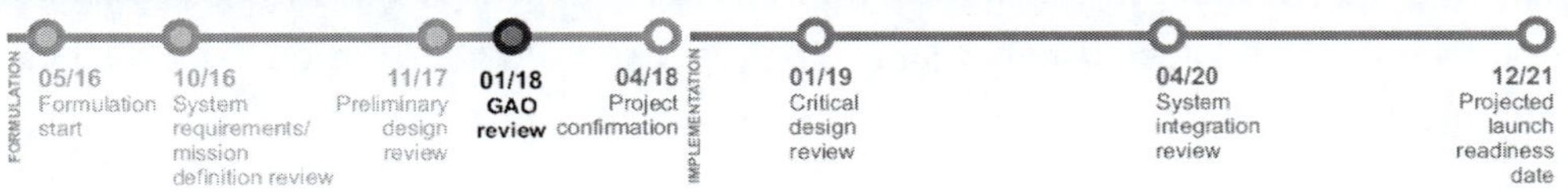

PROJECT INFORMATION

NASA Lead Center: **Goddard Space Flight Center**

International Partner: **None**

Launch Location: **Vandenberg Air Force Base, CA**

Launch Vehicle: **TBD**

Mission Duration: **12 months**

Requirement Derived from: **Consolidated Appropriations Act, 2016**

Budget Portfolio: **Space Technology, Research and Development**

PROJECT SUMMARY

The Restore-L project is no longer working to the preliminary cost and schedule estimates that NASA approved when it entered the preliminary design phase because the Space Technology Mission Directorate's proposed budget does not allow the project to execute to that plan. The mission directorate asked the project to evaluate two scenarios based on future funding possibilities. The project is to present a plan based on these scenarios to the mission directorate prior to the project establishing its cost and schedule baselines at its confirmation review in April 2018. The project reported that it does not use any new, critical technologies and that all six heritage technologies are modified from prior projects to fit the Restore-L design. The project matured five of these six technologies to the level recommended by best practices at its November 2017 preliminary design review. The technology that was not mature was a replacement for the prior vision navigation system, which did not meet requirements and the vendor was unable to resolve the issue. Project officials stated that they have a plan in place to mature the new system by December 2018.

PROJECT BUDGET INFORMATION
percent of current portfolio for fiscal year 2019

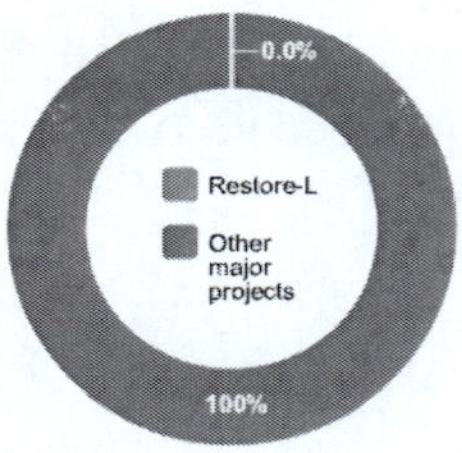

TOTAL FUNDED TO DATE
then-year dollars in millions

PRELIMINARY COST[a]
then-year dollars in millions

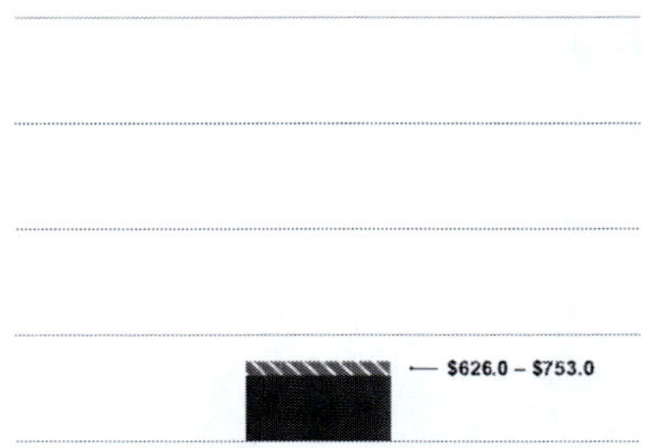

[a]*This estimate is preliminary, as the project is in formulation and there is uncertainty regarding the costs associated with the design options being explored. NASA uses these estimates for planning purposes.*

PRELIMINARY SCHEDULE

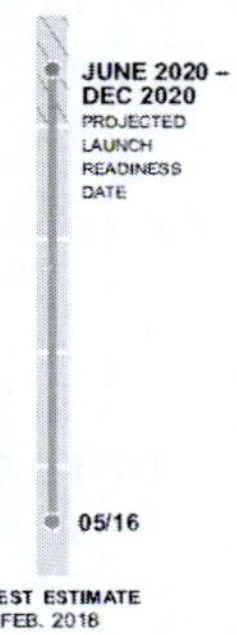

Cost and Schedule Status

The Restore-L project is no longer working to the preliminary cost and schedule estimates that NASA approved when it entered the preliminary design phase because the Space Technology Mission Directorate's proposed budget does not allow the project to execute to that plan. In April 2017, NASA set a preliminary cost estimate range of $626 million to $753 million with a projected launch readiness date between June and December 2020. However, the funding profile the Space Technology Mission Directorate proposed for future years will not allow the project to maintain a launch in 2020. As a result, the mission directorate asked the project to evaluate two scenarios based on future funding possibilities. According to NASA officials, one scenario assumes $130 million in project funding per fiscal year through the updated preliminary launch date for on-orbit demonstration. The other assumes $130 million for fiscal year 2018 followed by about $45 million for each of the fiscal years 2019 through 2024—focusing on technology development for industry use. The project is to present a plan based on these scenarios to the mission directorate prior to project confirmation review—the point at which the project will formally establish its cost and schedule baselines. The project has delayed its confirmation review from September 2017 to April 2018, in part to allow for more time to evaluate the scenarios. As of December 2017, the project's notional schedule is for a launch readiness date in December 2021, 1 year later than planned when the project entered the preliminary design phase. This notional schedule is based on various funding scenarios the project could face, and provides additional schedule reserve.

Technology and Design

The Restore-L project matured five of its six technologies to the level recommended by best practices at its preliminary design review in November 2017. Best practices recommend maturing technologies to a technology readiness level 6 by the project's preliminary design review to help minimize risks for space systems entering product development. The technology that was not mature at the review was a replacement vision navigation system that is new to the project. The prior vision navigation system did not meet requirements and the vendor was unable to resolve the issue. The project's independent review board stated that Restore-L needed a comprehensive plan to mature the technology at the review and project officials said they have a plan in place to mature the new system to a technology readiness level 6 by December 2018. The project reported that it does not use any new, critical technologies and that all 6 heritage technologies are modified from prior projects to fit the Restore-L design. Project officials said that although Restore-L uses all heritage technologies, the technologies will be used together in a new way to complete the mission's requirements.

The robot system is driving the project's schedule and the project has had to redesign system components, which consumed schedule reserve and put pressure on the project's preliminary launch schedule of December 2020. The robot system, which includes the spacecraft's robotic arm and associated components, will be used to capture the Landsat 7 spacecraft and will then employ tools to access the Landsat 7's propellant system, transfer propellant, and seal the system prior to releasing the spacecraft. The resolver pre-amp assembly—a component used to amplify signals to the robotic arm—is undergoing redesign to address signal distortion concerns. Another element, the robotic electronics units used to control the arm, is being redesigned to simplify and reduce the number of components in order to reduce schedule risk during integration and test. This redesign consumed schedule reserve before the project began working to a December 2021 notional schedule.

Other Issues to be Monitored

The project's current notional schedule would delay Restore-L's launch outside the servicing window for Landsat 7. Specifically, the agreed upon servicing window is January through September 2021, which the notional project launch schedule in December 2021 does not meet. However, that window may be extended if NASA reimburses the U.S. Geological Survey for additional costs to continue Landsat 7 operations.

Project Office Comments

Restore-L project officials provided technical comments on a draft of this assessment, which were incorporated as appropriate.

Space Launch System

The Space Launch System (SLS) is intended to be NASA's first human-rated heavy-lift launch vehicle since the Saturn V was developed for the Apollo program. SLS is planned to launch NASA's Orion spacecraft and other systems on missions between the Earth and Moon and to enable deep-space missions, including Mars. NASA is designing SLS to provide an initial lift capacity of 70 metric tons to low-Earth orbit, and be evolvable to 130 metric tons, enabling deep space missions. The 70-metric-ton capability will include a core stage, powered by four RS-25 engines, and two five-segment boosters. The 130-metric-ton capability will use a new upper stage and evolved boosters.

Source: NASA | GAO-18-280SP

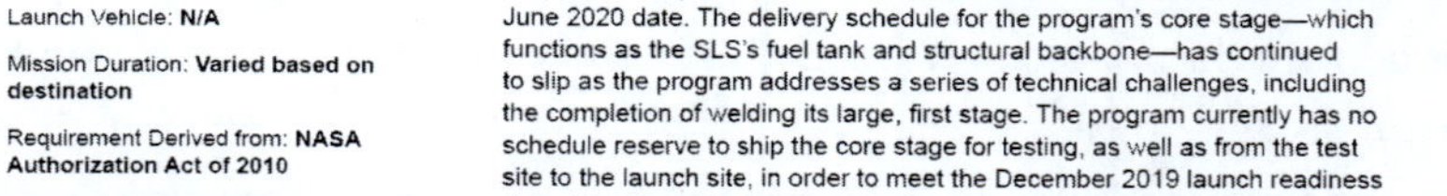

PROJECT INFORMATION

NASA Lead Center: **Marshall Space Flight Center**

International Partner: **None**

Launch Location: **Kennedy Space Center, FL**

Launch Vehicle: **N/A**

Mission Duration: **Varied based on destination**

Requirement Derived from: **NASA Authorization Act of 2010**

Budget Portfolio: **Exploration, Exploration Systems Development**

PROJECT SUMMARY

In December 2017, NASA announced a new schedule and cost estimate for the SLS program, after determining that the November 2018 launch readiness date for Exploration Mission-1 (EM-1) was no longer feasible. NASA is now planning to an internal launch readiness date of December 2019, with 6 months of schedule reserve to a June 2020 date. The life-cycle cost estimate for SLS is approximately $9.7 billion to the December 2019 date and $9.8 billion to the June 2020 date. The delivery schedule for the program's core stage—which functions as the SLS's fuel tank and structural backbone—has continued to slip as the program addresses a series of technical challenges, including the completion of welding its large, first stage. The program currently has no schedule reserve to ship the core stage for testing, as well as from the test site to the launch site, in order to meet the December 2019 launch readiness date. Should problems or issues arise during testing, they will likely affect the enterprise integration and test schedule.

PROJECT BUDGET INFORMATION
percent of current portfolio for fiscal year 2019

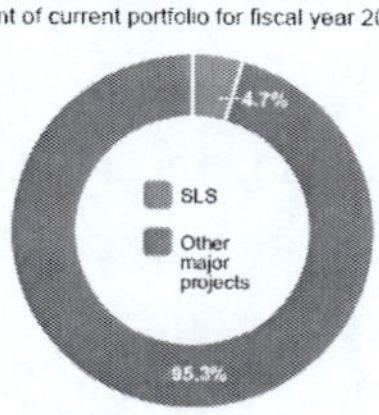

ESTIMATED FUNDING NEEDED
then-year dollars in millions

COST PERFORMANCE
then-year dollars in millions

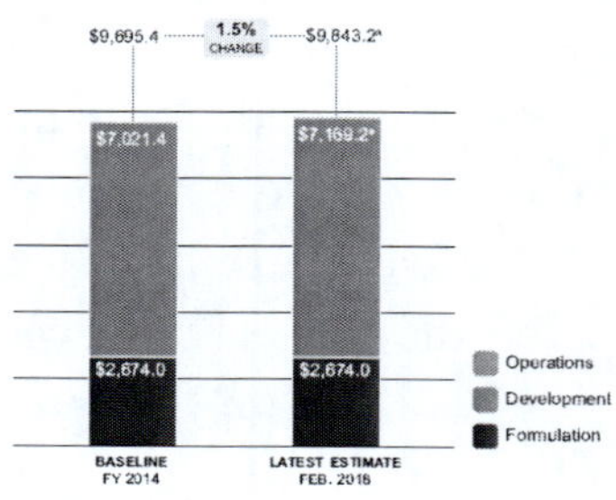

*Assumes June 2020 launch date.

SCHEDULE PERFORMANCE

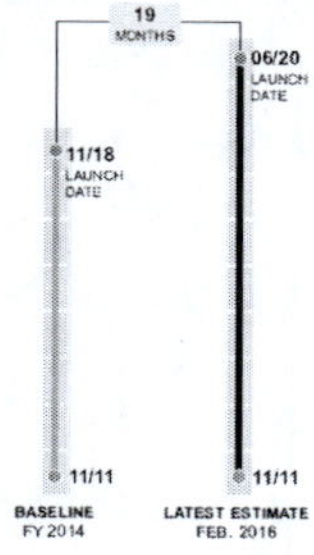

Cost and Schedule Status

In December 2017, NASA announced a new schedule and cost estimate for the SLS program, after determining that the November 2018 launch readiness date for Exploration Mission-1 (EM-1) was no longer feasible. In April 2017, we found that the date was likely unachievable for all three human paceflight programs—Exploration Ground Systems, SLS, and Orion Multi-Purpose Crew Vehicle—due to technical challenges continuing to cause schedule delays and the programs having little to no schedule reserve to the EM-1 date, meaning they would have to complete all remaining work with little margin for error for unexpected challenges that could arise.[48] The new internal launch readiness date for EM-1 is now December 2019, but NASA has also allocated 6 months of schedule reserve to June 2020 for possible manufacturing and production schedule risks. This represents a delay of 13-19 months for EM-1. The life-cycle cost estimate for SLS is about $9.7 billion to the December 2019 date and $9.8 billion to the June 2020 date, or 0.4 to 1.5 percent above the project's committed baseline.

Technology, Design, and Manufacturing

The delivery schedule for the program's core stage—which functions as the SLS's fuel tank and structural backbone—has slipped 14 months in the last year to May 2019, due in large part to production issues that have delayed completion of the core stage element. According to program officials, the liquid hydrogen tank has proven difficult to weld to specifications because it is thicker than metals that have been used by industry in the past. Without a history of similar efforts to compare to, officials explained that they have to check and test every potential anomaly with increased scrutiny than if they were producing something already used throughout the aerospace industry. In addition, according to officials, because this is the first flight article, the contractors are performing extensive testing to ensure that their production processes and methods are resulting in end items that meet or exceed specifications so that the next unit can be produced more quickly.

In addition to these issues, the program has faced work slowdowns and stoppages due to severe weather. In February 2017, a tornado hit parts of the Michoud Assembly Facility in Louisiana, where the core stage is being produced. While work resumed shortly after the storm, according to program officials, it did so at less than full capacity and delayed the program by a total of about 2 months. According to officials, repair work is complete in SLS manufacturing areas and should not further impact production work at the facility.

---

[48] GAO-17-414.

Integration and Test

The program has no schedule reserve through delivery of the core stage to Kennedy Space Center for the December 2019 launch date. In addition to completing production of flight and test articles, the program has to integrate the engines to the core stage and ship them to Stennis Space Center for a green run test. During this test, the core stage is fueled and the four main engines fired for about 500 seconds for the first time. This test will stress the flight components as well as the ground equipment used for the test. For example, according to program officials, 7 of the 12 cryogenic fluid pumps at Stennis Space Center must work together in order to fuel the vehicle. Should issues arise, the program will likely need additional time to assess and mitigate difficulties or glitches, which would likely affect delivery to Kennedy Space Center and could delay the enterprise integration and test schedule.

Project Office Comments

In commenting on a draft of this assessment, SLS program officials stated that a vast majority of SLS development and production work is on track for EM-1 and that work is also proceeding for the first crewed flight—EM-2. In addition, officials state that the program is successfully working through first-time production issues that are not unprecedented for a program of this scope and ambition. Program officials also provided technical comments, which were incorporated as appropriate.

Space Network Ground Segment Sustainment

The Space Network Ground Segment Sustainment (SGSS) project plans to develop and deliver a new ground system for one Space Network site. The Space Network provides essential communications and tracking services to NASA and non-NASA missions. Existing systems, based on 1980s technology, are increasingly obsolete and unsustainable. The new ground system will include updated systems, software, and equipment that will allow the Space Network to continue to provide critical communications services for the next several decades. The Space Network is managed by the Space Communication and Navigation (SCaN) program.

Source: NASA | GAO-18-280SP

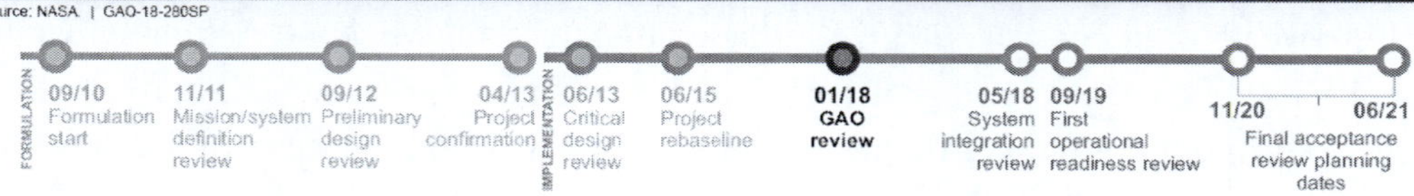

PROJECT INFORMATION

NASA Lead Center: **Goddard Space Flight Center**

International Partner: **None**

Launch Location: **N/A**

Launch Vehicle: **N/A**

Mission Duration: **25 years with periodic, required upgrades to hardware and software**

Requirement Derived from: **March 2008 Space Network modernization concept study**

Budget Portfolio: **Space Operations, Space and Flight Support**

PROJECT SUMMARY

In 2017, the SGSS project experienced additional cost growth and schedule delays beyond the rebaselined cost and schedule estimate NASA set in 2015. Since the 2015 rebaseline, the project's total costs increased by at least $112.9 million, from $1,207.9 million to $1,320.8 million even as the scope decreased from nine terminals at three Space Network sites to six terminals at one site. According to the fiscal year 2019 budget request, NASA plans to conduct an independent review in 2018 to inform a decision on whether to continue the project. If the project continues, additional cost growth is expected as the latest cost estimate covers only through the initial operational readiness review, currently planned for September 2019. Project officials attributed the cost growth and delays to an incomplete understanding of requirements by the contractor, which led to poor contractor plans and late design changes. Contractor performance has been a concern for the project, but project officials stated that contractor performance improved in fiscal year 2017. The project delivered the final software increment in May 2017 and the software is now in integration and testing.

PROJECT BUDGET INFORMATION
percent of current portfolio for fiscal year 2019

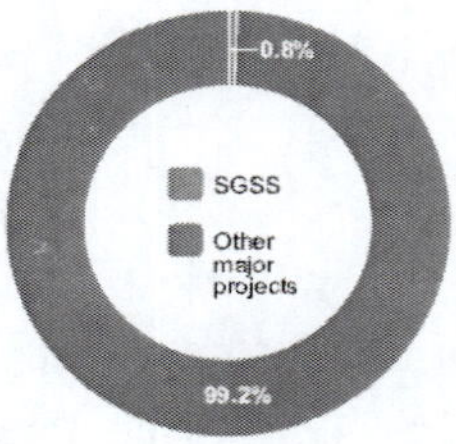

ESTIMATED FUNDING NEEDED
then-year dollars in millions

COST PERFORMANCE
then-year dollars in millions

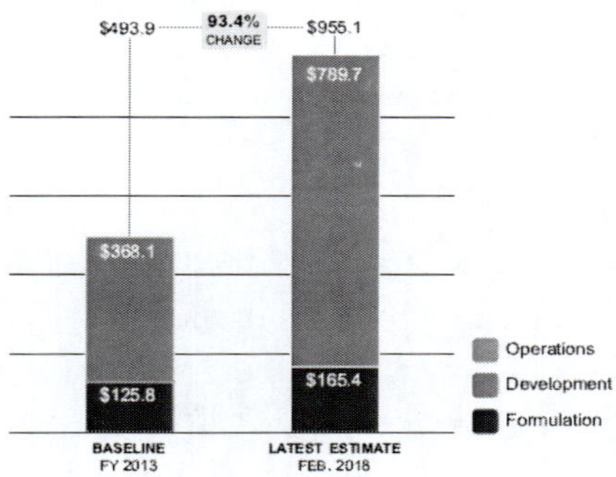

Note: The SGSS project has received an additional $365.7 million from Space Network users outside of NASA.

SCHEDULE PERFORMANCE

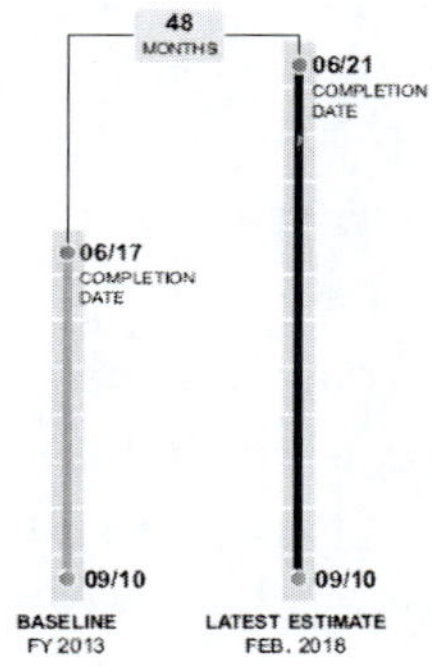

Cost and Schedule Status

In 2017, the SGSS project experienced additional cost growth and schedule delays beyond the cost and schedule rebaseline NASA set for it in 2015. In 2016, SGSS was reclassified as a hybrid sustainment project instead of a major project, which means that NASA is no longer measuring the project against its cost and schedule baseline, among other oversight implications. However, according to officials, the project is still required to provide quarterly and annual status reports to NASA on their cost and schedule. Since the 2015 rebaseline, however, the project's total costs increased by at least $112.9 million, from $1,207.9 million to $1,320.8 million, even as the scope decreased from upgrading nine terminals at three Space Network sites to six terminals at one site. The SGSS project presented its new cost and schedule estimate to NASA headquarters in August 2017, which NASA approved up through the initial operational readiness review, currently planned for September 2019.

Additional cost growth is expected for SGSS when NASA revisits project costs through future budget cycles. According to the fiscal year 2019 budget request, an independent review of the project is underway to determine whether the project will continue and was initiated due to budget challenges. In the event the project continues, officials stated the project plans to evaluate various funding scenarios, including continuing but receiving less funding than requested. Project officials attributed the cost growth and delays to an incomplete understanding of requirements by the contractor, which led to poor contractor plans and late design changes, but project management has been a challenge as well. For example, the project has historically struggled to manage contractor performance and has faced staffing shortfalls in key areas, such as systems engineering and business management. However, as of January 2018, NASA officials stated that these workforce issues had largely been resolved because some positions were filled by reassigning staff to these areas and others are no longer needed. The project delivered the final software increment in May 2017 and the software is now in integration and testing. The project is working toward the systems integration review in May 2018.

Contractor

Contractor performance has been an ongoing concern for the project, but project officials stated that performance improved in fiscal year 2017. The SGSS project attributes most of its cost and schedule growth to the contractor underestimating the scope of the development effort. The contractor moved a large portion of the software's required functionality into the last software development increment after experiencing problems with earlier increments. Project officials stated the contractor did not fully understand the requirements, technical planning was inadequate, and the contractor's planning did not account for resolving software defects. In response to the issues the project has faced, the SGSS project has taken action to try to address contractor performance problems. For example, the project worked with the contractor to develop new, more reliable cost and schedule estimates both for 2017 and beyond. The new schedule focuses on discrete tasks, which enables better tracking and measurability. In addition, the project managers for the project and the contractor were replaced. Project officials stated that contractor performance improved in fiscal year 2017 as the project successfully completed most of the plan it created for fiscal year 2017, which included delivery of the final software increment to integration and testing in May 2017.

Integration and Test

The project is tracking several risks to completing integration and testing—a period where problems are commonly found and schedules tend to slip—on schedule. For example, the project is tracking a risk that Space Network and mission partner users may be unavailable for testing, which could slow progress. The project is also tracking the number of software defects as a risk. While the project and the contractor's models generally agree on defect resolution timeframes, officials said there is a risk that defects could arise that would take longer than planned to fix. Additionally, the project is continuing work to demonstrate end-to-end performance and stability of the SGSS system. As part of the fiscal year 2017 plan, the project established five operational scenarios as measures of performance and stability. The project completed four of the operational scenarios, but was unable to complete the fifth due to a problem with third-party software. The project is working with the software vendor to resolve the issue.

Project Office Comments

In commenting on a draft of this assessment, project officials noted the contractor performed on plan for fiscal year 2017 and completed all the milestones within the critical path. The project also provided technical comments, which were incorporated as appropriate.

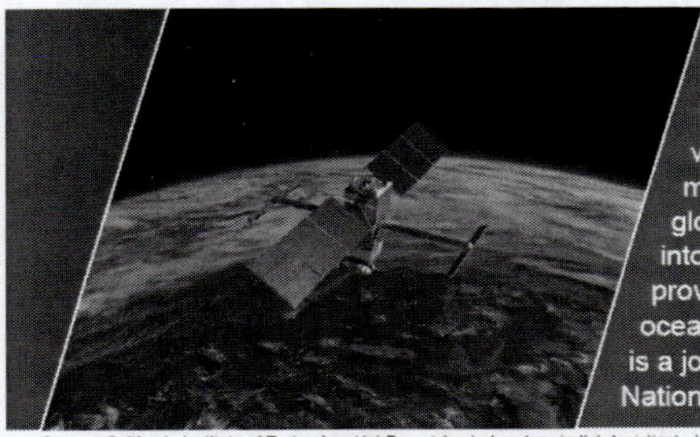

Surface Water and Ocean Topography

The Surface Water and Ocean Topography (SWOT) mission will use its wide-swath radar altimetry technology to take repeated high-resolution measurements of the world's oceans and freshwater bodies to develop a global survey. This survey will make it possible to estimate water discharge into rivers more accurately, and help improve flood prediction. It will also provide global measurements of ocean surface topography and variations in ocean currents, which will help improve weather and climate predictions. SWOT is a joint project between NASA and the French Space Agency—the Centre National d'Etudes Spatiales (CNES).

Source: California Institute of Technology/Jet Propulsion Laboratory (artist depiction). | GAO-18-280SP

FORMULATION

- 11/12 Formulation start
- 05/14 System requirements/ mission definition review
- 04/16 Preliminary design review
- 05/16 Project confirmation

IMPLEMENTATION

- **01/18 GAO review**
- 02/18 Critical design review
- 07/19 System integration review
- 04/21 Manifested launch readiness date
- 04/22 Committed launch readiness date

PROJECT INFORMATION

NASA Lead Center: **Jet Propulsion Laboratory**

International Partners: **Centre National d'Etudes Spatiales (France), Canadian Space Agency (Canada), United Kingdom Space Agency (United Kingdom)**

Launch Location: **Vandenberg Air Force Base, CA**

Launch Vehicle: **Falcon 9**

Mission Duration: **3 years**

Requirement Derived from: **2007 Earth Science Decadal survey**

Budget Portfolio: **Science, Earth Science**

PROJECT SUMMARY

The SWOT project continues to operate within its cost baseline and plans to launch in April 2021, a year earlier than its committed schedule baseline, despite encountering technical issues with its main instrument—the Ka-Band Radar Interferometer (KaRIn). An engineering model of the KaRIn high voltage power supply was damaged during testing due to three electrical arcing events that were caused by manufacturing and handling of parts, and weaknesses in the packaging design. In addition, the project augmented the main beams that connect the instrument antennas to the spacecraft, which resulted in delays to releasing design drawings. The project discovered signal interference off the main beams that could distort KaRIn's science measurements. Project officials said they planned to release about 90 percent of design drawings at the planned February 2018 critical design review, which is a best practice.

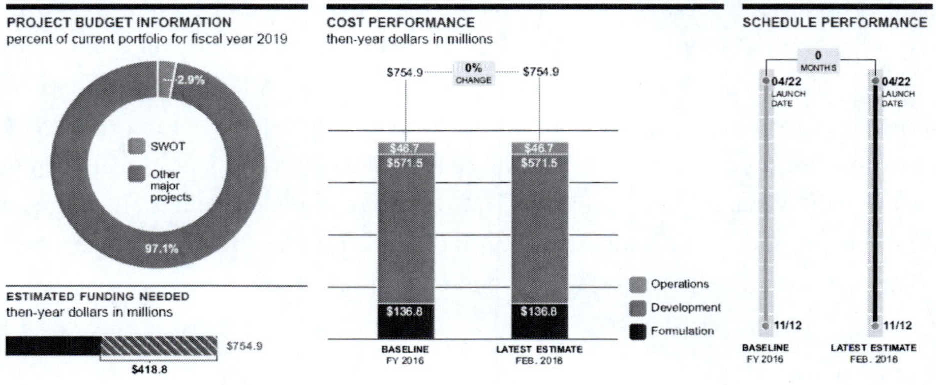

Cost and Schedule Status

The SWOT project continues to operate within its cost baseline and plans to launch in April 2021, a year earlier than its committed schedule baseline. The project is holding schedule reserves consistent with the level required by Jet Propulsion Laboratory policy, but did not meet cost reserve requirements in fiscal year 2017 due to costs associated with risk reduction activities and to stay on schedule. For example, project officials said they used cost reserves to build higher-fidelity engineering models and to secure a higher than planned workforce to execute fiscal year 2017 tasks. However, the project was able to reconstitute some of its cost reserves due to lower- than-expected launch vehicle costs and is meeting cost reserve requirements for fiscal year 2018 as of January 2018. The project is progressing toward its critical design review scheduled for February 2018.

Technology and Design

The project's primary instrument—the Ka-Band Radar Interferometer (KaRIn)—continues to be the most complicated development effort and has experienced technical issues that have had cost and schedule implications. During testing, an engineering model of the KaRIn high voltage power supply (HVPS) was damaged due to three electrical arcing events that were caused by manufacturing and handling of parts as well as weaknesses in the packaging design. To mitigate these events, the project repaired the electrical board within the engineering model with temporary fixes to support upcoming testing and built a second electrical board with enhanced handling procedures for a key thermal vacuum test in February 2018. As a result, HVPS development is driving the project's schedule.

To maintain its current schedule, the project is building the HVPS flight model in parallel to its ongoing testing of the engineering model. The project determined this approach was less risky than delaying flight model development until engineering model testing is complete. The project is tracking a risk, however, that development of the flight model concurrently may cause late rework.

In addition, the project augmented the main beams that connect the instrument antennas to the spacecraft, which resulted in delays to releasing design drawings. The project discovered signal interference off the main beams that could distort KaRIn's science measurements. To mitigate this risk, the project added a reflective plate to the main beam to minimize signal distortion and successfully held a mechanical critical design review with the modified main beam design. As a result, however, the project is 3 months behind schedule in releasing drawings for the modified design and is working with the Jet Propulsion Laboratory to secure additional staff to accelerate drawing production.

As of January 2018, SWOT released 87 percent of its expected drawings and project officials said they plan to release about 90 percent of design drawings at critical design review. Best practices show that releasing at least 90 percent of design drawings by this review can decrease a project's risk of cost growth and schedule delays.

Other Issues to Be Monitored

After the project experienced challenges with one of the tools—AirSWOT—that it plans to use to help understand the data returned from the KaRIn instrument once SWOT is in orbit, the project selected two additional methods to augment ocean measurements. The project determined that the performance of AirSWOT, which is an airborne sensor, was likely to be insufficient and that heavy wave activity when collecting ocean data was the root cause of measurement abnormalities. The project selected an airborne laser-based remote sensor along with a network of underwater gliders to augment the measurements from in-orbit satellites that were planned with AirSWOT. This approach successfully passed a technical peer review in December 2017 and the project plans to complete additional experiments using the selected approach to inform its calibration and validation plan.

Project Office Comments

In commenting on a draft of this assessment, project officials stated that SWOT is a challenging mission making a first of a kind measurement of global surface water. Officials also stated the project has been advancing the heritage systems development allowing the project to focus on the challenging KaRIn development. The project is completing thorough testing of engineering models along with a series of technical reviews in preparation for the flight model development. The project's focus remains on systematically rectifying technical issues while devising workarounds to maintain the overall milestones. SWOT officials also provided technical comments, which were incorporated as appropriate.

Transiting Exoplanet Survey Satellite

The Transiting Exoplanet Survey Satellite (TESS) will use four identical, wide field-of-view cameras to conduct the first extensive survey of the sky from space for transiting exoplanets—or planets in other solar systems. The mission's goal is to discover these exoplanets during transit, the time when the planet's orbit carries it in front of its star as viewed from Earth. The project plans to discover rocky and potentially habitable Earth-sized and super-Earth planets orbiting nearby bright stars for further evaluation through ground- and space-based observations by other missions, such as the James Webb Space Telescope.

Source: NASA. | GAO-18-280SP

FORMULATION: 04/13 Formulation start; 02/14 System requirements/ mission definition review; 09/14 Preliminary design review; 10/14 Project confirmation

IMPLEMENTATION: 08/15 Critical design review; 07/17 System integration review; **01/18 GAO review**; 04/18 Manifested launch readiness date; 06/18 Committed launch readiness date

PROJECT INFORMATION

NASA Lead Center: **Goddard Space Flight Center**

International Partner: **None**

Launch Location: **Cape Canaveral Air Force Station, FL**

Launch Vehicle: **Falcon 9**

Mission Duration: **2 years**

Requirement Derived from: **2010 Astrophysics Decadal Survey**

Budget Portfolio: **Science, Astrophysics**

PROJECT SUMMARY

The TESS project launched on April 18, 2018, before its committed launch date and within its cost baseline. The project expected to launch in March 2018 but did not meet this date due to certification delays with the SpaceX Falcon 9 upgrade known as Block 4. To cover the 1-month delay, NASA shifted funding previously budgeted for the operations phase to the development phase. The project did not have headquarters-held reserves to cover a launch past March 2018 because NASA reallocated $15 million of TESS's headquarters-held reserves to another astrophysics project. The TESS project completed environmental testing and conducted additional testing on its spare camera at temperatures seen in space to better understand expected camera performance on orbit. During thermal testing, the project found that the substance attaching the lenses to the camera barrel places pressure on the lenses and causes the cameras' focus to shift slightly. However, in June 2017, NASA directed the project to integrate the cameras because they are still expected to meet TESS's top-level science requirements even with the anomaly.

PROJECT BUDGET INFORMATION
percent of current portfolio for fiscal year 2019

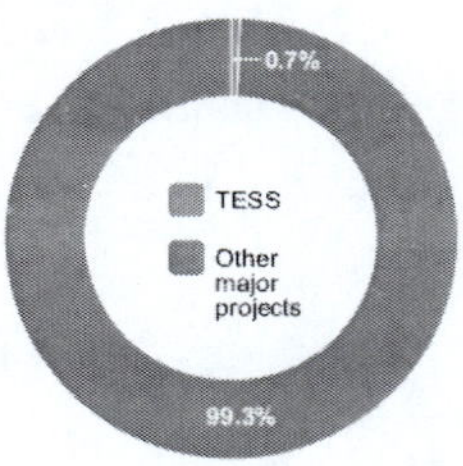

ESTIMATED FUNDING NEEDED
then-year dollars in millions

COST PERFORMANCE
then-year dollars in millions

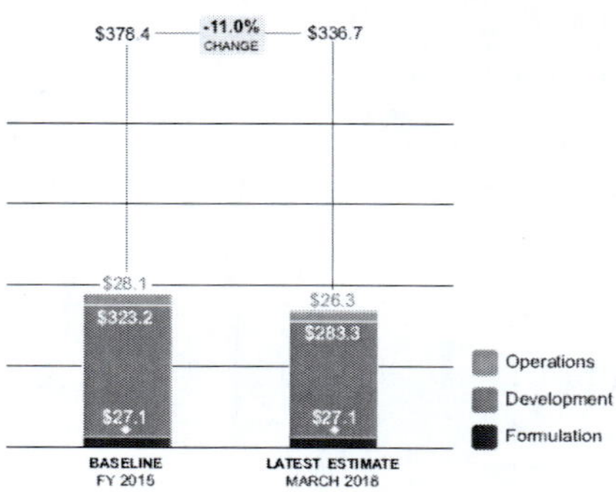

SCHEDULE PERFORMANCE

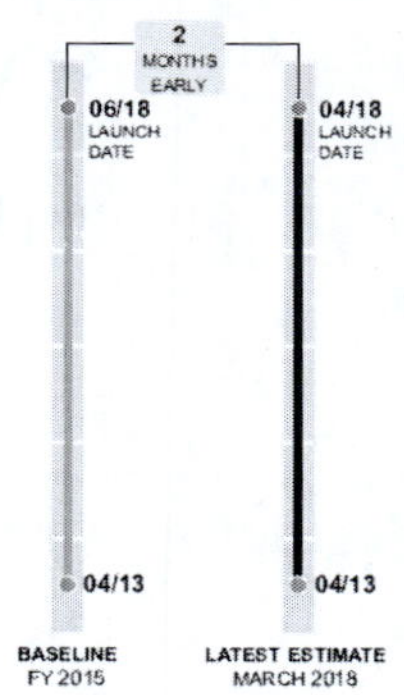

Cost and Schedule Status

The TESS project launched on April 18, 2018—2 months early and within its cost baseline. The project had planned to launch in March 2018, but the launch date was moved by a month due to delays related to its launch vehicle. At the key decision review prior to the project entering the operations and sustainment phase, NASA approved shifting $1.9 million previously budgeted for the operations phase to the development phase to cover the 1-month delay. The project did not have any headquarters-held cost reserves available to cover a launch past March 2018 because NASA reallocated $15 million of TESS's headquarters-held reserves to another astrophysics project, the Wide-Field Infrared Survey Telescope project. The project is holding cost and schedule reserves at the level required by NASA center policy.

Integration and Test

The project completed observatory environmental testing after overcoming spacecraft and instrument challenges to its cost and schedule. The project entered system assembly, integration and test—the phase where problems are most commonly found and schedules tend to slip—in August 2017. The project mitigated two risks related to its spacecraft. First, the project integrated its flight Ka-band transmitter in October 2017 after a 2-year delivery delay. The transmitter, essential for TESS as it transmits the mission data back to Earth, experienced continued contractor performance and manufacturing issues such as incorrectly placed parts. The project used schedule reserves to accommodate delays, installed an on-site representative to oversee the contractor's work, and developed a contingency plan to fly TESS with the transmitter's engineering development unit, which performed well in testing. Second, the project closed a risk that its slip rings—components within the solar array drive assembly necessary for power transmission—could be unsafe for flight. In May 2017, the project found damage to a small number of its slip rings, but later determined the slip rings were acceptable for flight after testing showed that they performed as expected.

The project conducted additional testing on its spare camera at temperatures experienced in space to better understand expected camera performance on orbit after finding that the focus of each camera shifts slightly at the temperatures in which TESS will operate. In thermal testing, the project found that the substance attaching the lenses to the camera barrel places pressure on the lenses and causes the cameras' focus to shift slightly at on-orbit temperatures. Prior to system integration review, NASA deemed the four cameras acceptable as they are expected to meet top-level science requirements even with the anomaly and in June 2017 directed the project to integrate the cameras. In July 2017, the project integrated the cameras with an alternative data handling unit, which powers the cameras and serves as the instrument data storage and processing computer. The project decided to use the alternative data handling unit because the unit originally designed for the project was behind schedule due to technical issues and contractor performance.

Launch

The project did not meet its expected March 2018 launch date due to certification delays for its launch vehicle, the SpaceX Falcon 9 upgrade known as Block 4. Certification is necessary because it will be the first time that a NASA instrument will launch on the Block 4 version of the vehicle. The TESS project expected that NASA's Launch Services Program would certify the Block 4 in September 2017, 7 months before TESS is scheduled to launch. However, NASA needed additional time to investigate the Falcon 9 second stage pressure vessel, which was involved in an anomaly that caused an explosion in September 2016. SpaceX also required extra time to meet NASA's Launch Services Program requirements.

Project Office Comments

TESS project officials provided technical comments on a draft of this assessment, which were incorporated as appropriate.

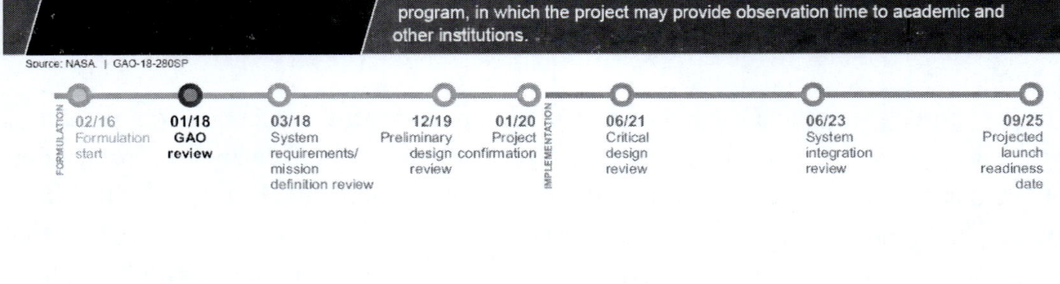

Source: NASA. | GAO-18-280SP

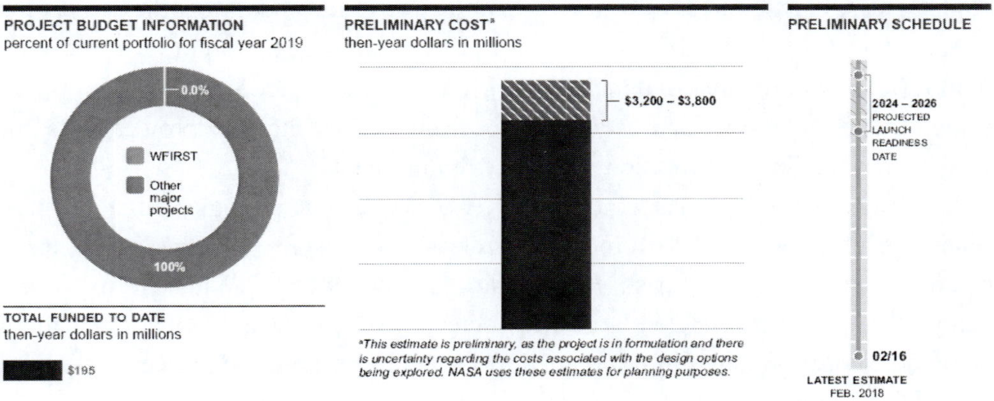

Cost and Schedule Status

The WFIRST project's preliminary cost and schedule estimates are currently under review as the project responds to findings from an October 2017 independent review conducted to ensure the mission's scope and required resources are well understood and executable. The review found that the project is not executable unless its mission scope is redesigned or its preliminary cost target is increased. NASA agreed with the study team's results and directed the project to reduce the cost and complexity of the design in order to maintain costs within the project's $3.2 billion preliminary cost target. In February 2018, the President's 2019 Budget Request proposed canceling the WFIRST project.

Design and Technology

In response to the independent review, the project developed a system design proposal that will be presented at its system requirements and mission definition review in March 2018 to reduce WFIRST cost and complexity. As of January 2018, the project was planning to reach the $3.2 billion cost target by treating the coronagraph—designed to perform high contrast imaging and spectroscopy of nearby exoplanets—as a technology demonstration as opposed to a science instrument. This, according to officials, allows the project to make a trade between performance and risk, and thereby avoid potential impacts to the project's cost and schedule. In addition, the project plans to review the capabilities of the Wide Field Instrument, which is intended to measure light from a billion galaxies and perform a survey of the inner Milky Way, as part of the review. NASA is also considering whether or not to maintain capability for WFIRST to be "starshade ready." A starshade is a device that is launched with or separately from an observatory and positioned between it and the star being observed to block out the starlight while allowing the light emitted by the planet through. NASA planned to review and formally approve the changes at the project's next key decision review, planned for April 2018, prior to entering the preliminary design and technology completion phase. Finally, in response to the independent review, NASA upgraded the risk classification of the project, which could increase project costs due to more system reliability and mission assurance requirements.

The WFIRST project has matured its two critical technologies to a technology readiness level 6, which helps to minimize risks for space systems when entering production, but is tracking a risk concerning the production of Wide Field Instrument detectors. Project officials stated large array detectors always pose fabrication challenges. To date, detector production has exceeded historical trends, but the project is completing ongoing tests to validate detector quality.

Developmental Partner

NASA is considering several potential contributions with various international partners, including the Canadian Space Agency, European Space Agency, France, Germany, and Japan, for elements of the Wide Field Instrument, coronagraph, and ground system. These contributions could potentially reduce the project's cost. NASA also expects to determine its international partners before the project enters the preliminary design and technology completion phase.

Project Office Comments

WFIRST project officials provided technical comments on a draft of this assessment, which were incorporated as appropriate.

AGENCY COMMENTS AND OUR EVALUATION

We provided a draft of this chapter to NASA for its review and comment. In its written response, NASA generally agreed with our findings and stated that the chapter provides NASA with a valued independent perspective on its major acquisitions. NASA also noted that we incorporated technology demonstration projects into the audit this year, and that it is important to remember that the purpose of these activities is to mature specific new technologies to a technology readiness level 6 or higher by the end of their demonstration phase in a relevant environment. As a result, it is NASA's view that our best practice to mature technologies to a technology readiness level 6 by the preliminary design review is not applicable to technology demonstration projects.

As discussed in the chapter, we included two technology demonstration projects—LCRD and Restore-L—in our assessment this year because these two projects each had a life-cycle cost estimate of over $250 million. In addition, we included these two projects in our analysis of NASA major projects attaining technology maturity at preliminary design review because both projects planned to mature their technologies prior to launch, not at the end of their demonstration phase in a relevant environment as NASA described in its letter. Therefore, these two projects are still susceptible to the same risks projects might experience if they fall short of the best practice of meeting a technical readiness level 6 by preliminary design review. These risks include subsequent technical problems that could result in cost growth and schedule delays. In the future, if other technology demonstration projects have a life-cycle cost estimate over $250 million and enter the portfolio of projects that we assess, we will continue to discuss project technology maturity goals for those specific projects with NASA. These discussions will inform our determination of whether we will compare those projects against our best practices on technology maturity.

NASA's written comments are reprinted in appendix VI. NASA also provided technical comments, which we incorporated as appropriate.

Contact points for our Offices of Congressional Relations and Public Affairs may be found on the last page of this chapter. GAO staff who made major contributions to this chapter at listed in appendix VII.

List of Committees

The Honorable Jerry Moran
Chairman

The Honorable Jeanne Shaheen
Ranking Member
Subcommittee on Commerce, Justice, Science, and Related Agencies
Committee on Appropriations
United States Senate

The Honorable Ted Cruz
Chairman

The Honorable Edward J. Markey
Ranking Member

Subcommittee on Space, Science, and Competitiveness Committee on Commerce, Science, and Transportation United States Senate

The Honorable John Culberson
Chairman

The Honorable José Serrano
Ranking Member
Subcommittee on Commerce,
Justice, Science, and Related Agencies
Committee on Appropriations
House of Representatives

The Honorable Brian Babin
Chairman

The Honorable Ami Bera Ranking Member
Subcommittee on Space
Committee on Science, Space, and Technology
House of Representatives

APPENDIX I: OBJECTIVES, SCOPE, AND METHODOLOGY

The objectives of our review were to assess (1) the cost and schedule performance of the National Aeronautics and Space Administration's (NASA) portfolio of major projects, (2) the maturity of technologies and stability of project designs at key points in the development process, and (3) the extent to which NASA is assessing future workforce capacity challenges that may affect its ability to manage its portfolio of projects. We also described the status and assessed the risks and challenges faced by NASA's 26 major projects, each with life-cycle costs more than $250 million. When NASA determines that a project has an estimated life- cycle cost of over $250 million, we include that project in our annual review up through launch or completion. We did not develop an assessment for the Asteroid Redirect Robotic Mission project because, as of June 2017, the project closed out most project development activities at the direction of NASA headquarters.

To respond to these objectives, we developed several standard data collection instruments (DCI). We developed multiple DCIs, which were completed by NASA's Office of the Chief Financial Officer, to gather data on each project's cost, schedule, funding, and notional 5-year budget. We used another DCI, which was completed by each project office, to gather data on project's technology and design maturity and development partners. The information available on individual projects depends on where a project is in its life cycle. For example, for projects in an early stage of development called formulation there are still unknowns about requirements, technology, and design. We also analyzed DCI data from prior reviews.

To assess the cost and schedule performance of NASA's major projects, we compared cost and schedule data as of February 2018 provided on DCIs by NASA for the 17 projects in the implementation phase during our review to previously established cost and schedule baselines.[49] The Commercial Crew Program has a tailored project life cycle and project management requirements, so it was excluded from this analysis. In addition, we assessed development cost and schedule performance for NASA's portfolios of major projects for each reporting period between 2009 and 2018 to examine longer-term trends. To determine cost performance, we compared the projects' baseline development costs and development costs as of February 2018. For the Orion program, we used a June 2017 cost estimate because NASA was revising the program's life-cycle cost estimate at the time of our review. For two projects — Gravity Recovery and Climate Experiment Follow-On (GRACE-FO) and Transiting Exoplanet Survey Satellite (TESS)—the cost performance is current as of March 2018 because NASA updated cost information after we received the cost DCI in February 2018. All cost information in this chapter is presented in nominal then-year dollars for consistency with budget data. Current baseline costs for all projects are adjusted to reflect the cost accounting structure in NASA's fiscal year 2009 budget estimates. For the fiscal year 2009 budget request, NASA changed its accounting practices from full-cost accounting to reporting only direct costs at the project level. To determine schedule performance, we compared the project's baseline launch readiness or completion date and launch readiness or completion date as of February 2018. For three projects—GRACE-FO, James Webb Space Telescope, and TESS—the schedule performance is current as of March 2018 because NASA provided an updated schedule after we received the schedule DCI in February 2018.

To assess technology maturity, we asked project officials to complete a DCI that provided the technology readiness levels of each of the project's critical and heritage technologies at various stages of project development including the preliminary design review. We did not verify or validate project office supplied data on the technology readiness level of technologies, or the classification of technologies as critical or heritage. For the 17 projects that had held a preliminary design review and identified critical or heritage technologies, we compared those levels against our technology maturity best practice and NASA policy on technology maturity to determine the extent to which the portfolio was meeting the criteria. Our work has shown that reaching a technology readiness level 6—which indicates that the representative prototype of the technology has

---

[49] For the purpose of this review, cost performance is defined as the percentage of total development cost growth over the development cost baseline.

been demonstrated in a relevant environment that simulates the harsh conditions of space—by the preliminary design review is the level of maturity needed to minimize risks for space systems entering product development. Originally developed by NASA, technology readiness levels are measured on a scale of one to nine, beginning with paper studies of a technology's feasibility and culminating with a technology fully integrated into a completed product. See appendix IV for the definitions of technology readiness levels. We compared this year's results against those in prior years to assess whether NASA was improving in this area. We did not assess technology maturity for those projects that had not yet reached the preliminary design review at the time of this assessment or for projects that reported no critical or heritage technologies. We also excluded 2009 from our analysis since the data were only for critical technologies and did not include heritage technologies. We compared the number of critical technologies being developed per project with those in prior years to determine how the number of critical technologies developed per project had changed. We also collected information on the use of heritage technologies in the projects; including what heritage technologies were being used; what effort was needed to modify the form, fit, and function of the technology for use in the new system; and whether the project considered the heritage technology as a risk to the project.

To assess design stability, we asked project officials to complete a DCI that provided the number of engineering drawings completed or projected for release by the preliminary and critical design reviews and as of our current assessment.[50] We did not verify or validate project office supplied data on the number of released and expected engineering drawings. However, we collected the project offices' rationale for cases where it appeared that only a small percentage of the expected drawings were completed by the time of the design reviews or where the project office reported significant growth in the number of drawings released after the critical design review. In accordance with best practices, projects were assessed as having achieved design stability if at least 90 percent of projected drawings were released by the critical design review. We omitted the Exploration Ground Systems program in this year's analysis and retroactively excluded the program from prior year analyses in 2016 and 2017 because program officials said they do not track design drawings as a management tool and we found that the prior data submissions were not inclusive all subsystems. We compared this year's results against those in prior years to assess whether NASA was improving in this area. For this year's assessment, 12 projects had held a critical design review and reported data on design drawings. We did not assess the design stability for those projects that had not yet reached the critical design review at the time of this assessment. To assess project technical margins, we gathered project mass and power information using a DCI and compared it against NASA requirements. We omitted the

[50] In our calculation for the percentage of total number of drawings projected for release, we used the number of drawings released at the critical design review as a fraction of the total number of drawings projected, including where a growth in drawings occurred. Therefore, the denominator in the calculation may have been larger than what was projected at the critical design review. We believe that this more accurately reflected the design stability of the project.

Exploration Ground Systems, Space Network Ground Segment Sustainment, and Space Launch System as those projects do not contain spacecraft. We excluded the Orion program because it does not have applicable metrics. To assess completion of project validation and verification plans, we asked project officials to complete a DCI that provided data on whether a plan was completed by the critical design review.

To assess the extent to which NASA is assessing future workforce capacity challenges that may affect its ability to manage its portfolio of projects, we did the following:

- Analyzed NASA major project documents, such as monthly status reports, and interviewed project officials to determine if the projects were experiencing any workforce challenges. We also discussed the process of staffing Jet Propulsion Laboratory projects with officials within the NASA Management Office at the laboratory.
- Interviewed Mission Support Directorate and Office of the Chief Engineer officials responsible for agency-wide capability assessments to identify if workforce challenges were raised as a concern in these assessments. Through an initial review of documents and discussions with officials, we identified select technical areas that workforce capability gaps were identified in the assessments for additional review, including the technical capability areas of sensors and instruments, and guidance, navigation and control. We also selected the human capital, budget, and procurement business assessments because of the relation to the acquisitions process and acquisition workforce functions that could affect NASA's ability to manage its major projects. We reviewed the recommendations in each assessment and identified those specifically related to aspects of workforce capacity and human capital management. We interviewed relevant officials on the status of implementation plans for these recommendations—including NASA's efforts to improve strategic workforce planning and cross-center collaboration—and discussed the role these efforts would play in the broader implementation of NASA's operating model.
- Analyzed publicly available workforce data from NASA's Workforce Information Cubes for NASA (WICN) to identify workforce characteristics that could affect NASA's workforce capacity. WICN contains snapshots of workforce composition as of certain dates and are updated every two weeks by NASA's Shared Services Center. We obtained and analyzed data from WICN as of January 2018. Specifically, we examined (1) total employee staffing levels for the past 5 years, (2) age, (3) retirement eligibility, and (4) the average number of years staff stay past initial retirement eligibility by occupational category. To assess the reliability of the workforce data used in the review, we reviewed NASA's responses to questions on the database,

which included information on the source of the data, how NASA uses the data, and the reliability of the data; interviewed officials with knowledge of how the database is used at the agency and how the information is compiled; and reviewed the data for errors, such as by confirming correct sums and checking the consistency of workforce totals across the data. We concluded the workforce data were sufficiently reliable for purposes of this chapter.

Our work was performed primarily at NASA headquarters in Washington, D.C. In addition, we and related GAO engagement teams visited Goddard Space Flight Center in Greenbelt, Maryland; the Jet Propulsion Laboratory in Pasadena, California; Kennedy Space Center in Merritt Island, Florida; Johnson Space Center in Houston, Texas; and Marshall Space Flight Center in Huntsville, Alabama.

Project Profile Information on Each Individual Project Assessment

This year, we developed project assessments for the 26 projects in the portfolio with an estimated life-cycle cost greater than $250 million. For each project assessment we included a description of each project's objectives, information concerning the NASA center, and international partners involved in the project, if applicable, the project's cost and schedule performance, a schedule timeline identifying key project dates, and a brief narrative describing the current status of the project. We also included budget information, including the percentage of NASA's fiscal year 2019 budget for all major projects in the current portfolio that the project represents. The budget information is based on NASA's fiscal year 2019 budget. For projects in formulation, we included the total funding that has been allocated to the project since formulation start through the end of fiscal year 2017. For projects in implementation, we included the funding needed to be allocated for project completion or launch through the project's current life-cycle cost estimate. We also provided a detailed discussion of project challenges for selected projects as applicable.

To assess the cost and schedule changes of each project, we obtained data directly from NASA's Office of the Chief Financial Officer through our DCI. For the Commercial Crew program, we obtained data directly from the program on the total amount of funds obligated and the schedule. When applicable, we compared the level of cost and schedule reserves held by the project to the level required by center policy. To determine the funding received to date for each of the projects in formulation we calculated the total funding allocated to the project since formulation start.[51] For projects in implementation, we calculated the funding needed for project launch or completion through its current life-cycle cost estimate.

[51] This does not include funds for studies prior to key decision point (KDP)-A, the start of formulation.

The project's timeline is based on acquisition cycle time, which is defined as the number of months between the project's start, or formulation start, and the projected or actual launch date. Formulation start generally refers to the initiation of a project; NASA refers to a project's start as key decision point (KDP)-A, or the beginning of the formulation phase. The preliminary design review typically occurs toward the end of the formulation phase, followed by a review at KDP-C, known as project confirmation, which allows the project to move into the implementation phase. The critical design review is generally held during the latter half of the final design and fabrication phase of implementation and demonstrates that the maturity of the design is appropriate to support continuing with the final design and fabrication phase. The manifested launch date is the launch date which the project is working toward, and when a launch vehicle is available to launch the project. This date is only a goal launch date for the project, not a commitment that they will launch on this date. The committed launch readiness date is determined through a launch readiness review that verifies that the launch system, spacecraft, and payloads are ready for launch. The implementation phase includes the operations of the mission and concludes with project disposal.

Project Challenges Discussion on Each Individual Project Assessment

To assess the status, risk, and challenges for each project, we submitted a DCI to each project office. In the DCI, we requested information on the maturity of critical and heritage technologies, the number of releasable design drawings at project milestones, and international partnerships.[52]

We also held interviews with representatives from all of the projects to discuss the information on the DCI. We then reviewed project documentation—including project plans, schedules, risk assessments, and major project review documentation—to corroborate any testimonial evidence we received in the interviews. These reviews led to identification of further challenges faced by NASA projects. The second page of our project assessments highlights key challenges facing that project that have or could affect project performance. For this year's report, we identified challenges across the projects we reviewed in the categories of launch, contractor, development partner, design, technology, schedule, and integration and test. These challenges do not represent an exhaustive or exclusive list and are based on our definitions and assessments, not those of NASA.

To supplement our analysis, we relied on our work over past years examining acquisition issues across multiple agencies. These reports cover such issues as contracting, program management, acquisition best practices, and cost estimating. We also have an extensive body of work related to challenges NASA has faced with specific system

[52] We did not collect this information for the Commercial Crew Program.

acquisitions, financial management, and cost estimating. This work provided the historical context and basis for large parts of the general observations we made about the projects we reviewed.

Data Limitations

NASA provided preliminary estimated life-cycle cost ranges and associated schedules for the eight projects that had not yet entered implementation, which are generally established at KDP-B. NASA formally establishes cost and schedule baselines, committing itself to cost and schedule targets for a project with a specific and aligned set of planned mission objectives, at KDP-C, which follows a preliminary design review. KDP-C reflects the life-cycle point where NASA approves a project to leave the formulation phase and enter into the implementation phase. NASA explained that preliminary estimates are generated for internal planning and fiscal year budgeting purposes at KDP-B, which occurs midstream in the formulation phase, and hence, are not considered a formal commitment by the agency on cost and schedule for the mission deliverables. Due to changes that occur to a project's scope and technologies between KDP-B and KDP-C, the estimates of project cost and schedule can be significantly altered between the two KDPs.

We conducted this performance audit from April 2017 to May 2018 in accordance with generally accepted government auditing standards. Those standards require that we plan and perform the audit to obtain sufficient, appropriate evidence to provide a reasonable basis for our findings and conclusions based on our audit objectives. We believe that the evidence obtained provides a reasonable basis for our findings and conclusions based on our audit objectives.

Appendix II: Major NASA Projects Assessed in GAO's 2018 Report

In 2018, we assessed 26 major NASA projects. Figure 11 shows the preliminary launch readiness data and cost estimates for projects in the formulation phase, and the current launch readiness dates and cost estimates for projects in the implementation phase.

Phase	Acronym	Project name	Preliminary launch readiness date	Preliminary cost estimate (in millions)
Formulation	DART	Double Asteroid Redirection Test	Dec 2020 – May 2021	$247.7 – $290.7
	Clipper	Europa Clipper	July 2023	$3,100 – $4,000
	LBFD	Low Boom Flight Demonstrator	March 2022	$390+
	Lucy	Lucy	Oct – Nov 2021	$914 – $984
	PACE	Plankton, Aerosol, Cloud, ocean Ecosystem	Aug 2022 – Apr 2023	$805 – $850
	Psyche	Psyche	August 2022	$907.3 – $957.3
	Restore-L	Restore-L	June – Dec 2020	$626 – $753
	WFIRST	Wide-Field Infrared Survey Telescope	2024 – 2026	$3,200 – $3,800

Phase	Acronym	Project name	Current launch readiness date	Current cost estimate (in millions)
Implementation	EGS	Exploration Ground Systems	June 2020	$3,230.7
	GRACE-FO	Gravity Recovery and Climate Experiment Follow-On	May 2018	$431.9
	ICESat-2	Ice, Cloud, and Land Elevation Satellite-2	October 2018	$1,063.6
	ICON	Ionospheric Connection Explorer	No earlier than June 2018	$252.7
	InSight	Interior Exploration using Seismic Investigations, Geodesy, and Heat Transport	May 2018	$828.9
	JWST	James Webb Space Telescope	May 2020	$8,825.4
	L9	Landsat 9	November 2021	$885.0
	LCRD	Laser Communications Relay Demonstration	November 2019	$262.7
	Mars 2020	Mars 2020	July 2020	$2,458.2
	NISAR	NASA Indian Space Research Organisation Synthetic Aperture Radar	September 2022	$896.9
	PSP	Parker Solar Probe[a]	August 2018	$1,548.2
	Orion	Orion Multi-Purpose Crew Vehicle[b]	April 2023	Under revision
	RBI	Radiation Budget Instrument (canceled)	December 2019	$304.8
	SGSS	Space Network Ground Segment Sustainment[c]	June 2021	$1,320.8
	SLS	Space Launch System	June 2020	$9,843.2
	SWOT	Surface Water and Ocean Topography	April 2022	$754.9
	TESS	Transiting Exoplanet Survey Satellite	April 2018	$336.7
	CCP	Commercial Crew Program[d]	Jan - Feb 2019	$6,968.4

Source: GAO analysis of National Aeronautics and Space Administration data. | GAO-18-280SP

Note: The life cycle for NASA space flight projects consists of two phases—formulation, which takes a project from concept to preliminary design, and implementation, which includes building, launching, and operating the system, among other activities. For projects in implementation, the current launch readiness date and cost estimate are the project's established cost and schedule baseline or the latest cost estimate and schedule if the project has experienced cost or schedule growth above the project's baseline.

[a]In May 2017, NASA renamed the Solar Probe Plus project as the Parker Solar Probe project.

[b]NASA officials said they are revising the Orion program's life-cycle cost estimate and expect to complete a new estimate in June 2018. The new cost is expected to be over the program's cost baseline of $11,283.5 million.

[c]In 2016, NASA reclassified the Space Network Ground Segment Sustainment (SGSS) as a hybrid sustainment effort, rather than a major project. A hybrid sustainment effort still includes development work. As a result, we continue to include SGSS in our assessment. The SGSS project reported costs are up through its first operational readiness review at the end of fiscal year 2019, and its schedule is under review.

[d]The launch readiness date for the Commercial Crew Program is for the certification reviews for Boeing and SpaceX. The Commercial Crew Program is implementing a tailored version of NASA's space flight project life cycle, but it is currently completing development activities typically associated with implementation.

Figure 11. Cost and Schedule of Major NASA Projects Assessed in GAO's 2018 Report by Phase.

APPENDIX III: MAJOR NASA PROJECTS REVIEWED IN GAO'S ANNUAL ASSESSMENTS

We have reviewed 55 major National Aeronautics and Space Administration (NASA) projects or programs since our initial review in 2009. See Figure 12 below for a list of projects included in our assessments from 2009 to 2017. These projects were not included in the 2018 review because they launched, were canceled, or launched but failed to reach orbit.

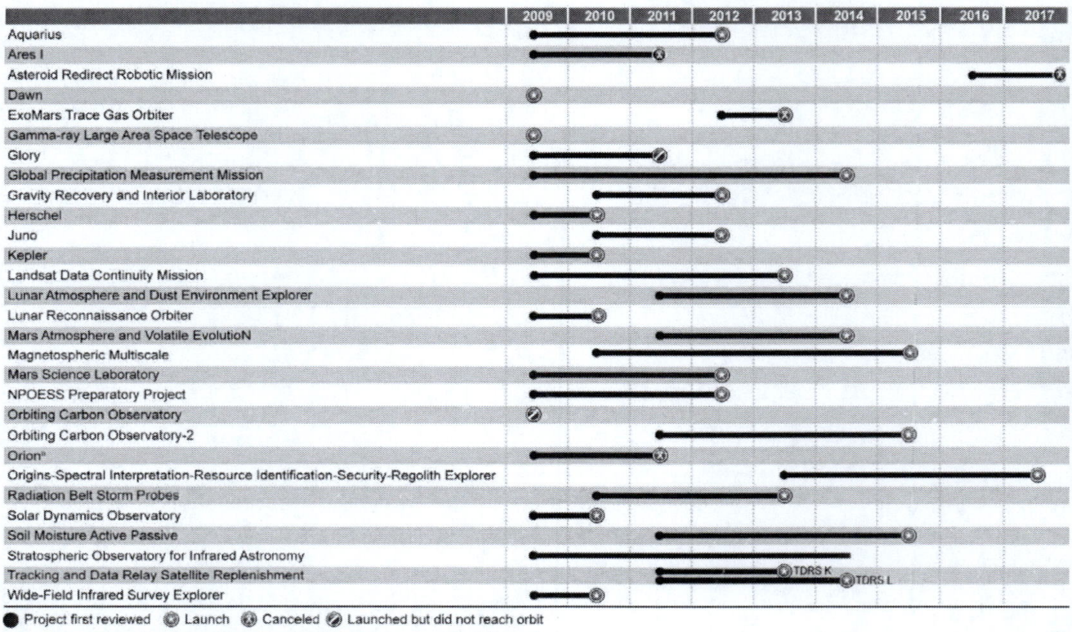

Source: GAO analysis of National Aeronautics and Space Administration data. | GAO-18-280SP

[a]In 2014, NASA adopted Orion as the common name for Orion MPCV; the project did not change. This Orion project stems from the original Orion project that was canceled in June 2011 when the Constellation program was canceled after facing significant technical and funding issues. During the closeout process for the Constellation program, NASA identified elements of the Ares I and Orion projects that would be transitioned for use on the new Space Launch System and Orion Multi-Purpose Crew Vehicle programs.

Figure 12. Major NASA Projects Reviewed in GAO's Annual Assessments from 2009-2017.

See Figure 13 below for a list of projects included in our 2018 assessment, including when the projects were first included in the review, and projects that were canceled during the review.

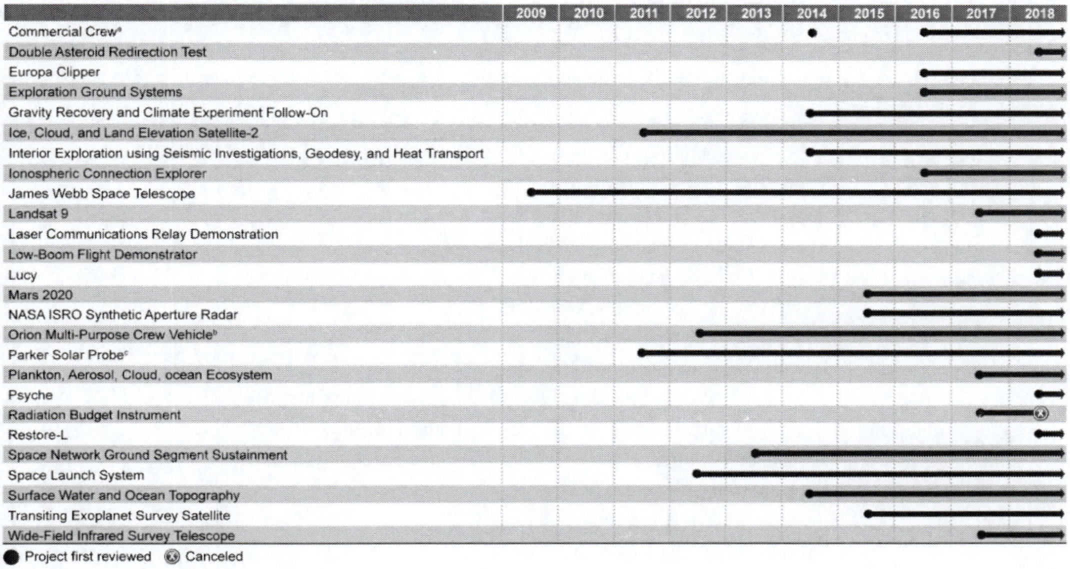

Source: GAO analysis of National Aeronautics and Space Administration data. | GAO-18-280SP

[a]A bid protest was filed on September 26, 2014, after NASA awarded Commercial Crew contracts. GAO issued a decision on the bid protest on January 5, 2015, which was after our review of projects had concluded; therefore, we excluded the Commercial Crew Program from the 2015 review.

[b]In 2014, NASA adopted Orion as the common name for Orion MPCV; the project did not change. This Orion project stems from the original Orion project that was canceled in June 2011 when the Constellation program was canceled after facing significant technical and funding issues. During the closeout process for the Constellation program, NASA identified elements of the Ares I and Orion projects that would be transitioned for use on the new Space Launch System and Orion Multi-Purpose Crew Vehicle programs.

[c]In May 2017, NASA renamed the Solar Probe Plus project as the Parker Solar Probe project.

Figure 13. Major NASA Projects Reviewed in GAO's 2018 Assessment.

APPENDIX IV: TECHNOLOGY READINESS LEVELS

Table 4. Characteristics of Technology Readiness Levels

Technology readiness level		Description	Hardware	Demonstration environment
1.	Basic principles observed and reported.	Lowest level of technology readiness. Scientific research begins to be translated into applied research and development. Examples might include paper studies of a technology's basic properties.	None (paper studies and analysis).	None.

Table 4. (Continued)

Technology readiness level		Description	Hardware	Demonstration environment
2.	Technology concept and/or application formulated.	Invention begins. Once basic principles are observed, practical applications can be invented. The application is speculative and there is no proof or detailed analysis to support the assumption. Examples are still limited to paper studies.	None (paper studies and analysis).	None.
3.	Analytical and experimental critical function and/or characteristic proof of concept.	Active research and development is initiated. This includes analytical studies and laboratory studies to physically validate analytical predictions of separate elements of the technology. Examples include components that are not yet integrated or representative.	Analytic studies and demonstration of nonscale individual components (pieces of subsystem).	Lab.
4.	Component and/or breadboard. Validation in laboratory environment.	Basic technological components are integrated to establish that the pieces will work together. This is relatively "low fidelity" compared to the eventual system. Examples include integration of ad-hoc hardware in a laboratory.	Low fidelity breadboard. Integration of nonscale components to show pieces will work together. Not fully functional or form or fit but representative of technically feasible approach suitable for flight articles.	Lab.
5.	Component and/or breadboard validation in relevant environment.	Fidelity of breadboard technology increases significantly. The basic technological components are integrated with reasonably realistic supporting elements so that the technology can be tested in a simulated environment. Examples include high-fidelity laboratory integration of components.	High-fidelity breadboard. Functionally equivalent but not necessarily form and/or fit (size, weight, materials, etc.). Should be approaching appropriate scale. May include integration of several components with reasonably realistic support elements/subsystems to demonstrate functionality.	Lab demonstrating functionality but not form and fit. May include flight demonstrating breadboard in surrogate aircraft. Technology ready for detailed design studies.

Technology readiness level		Description	Hardware	Demonstration environment
6.	System/subsystem model or prototype demonstration in a relevant environment.	Representative model or prototype system, which is well beyond the breadboard tested for technology readiness level 5, is tested in a relevant environment. Represents a major step up in a technology's demonstrated readiness. Examples include testing a prototype in a high-fidelity laboratory environment or in simulated realistic environment.	Prototype. Should be very close to form, fit, and function. Probably includes the integration of many new components and realistic supporting elements/subsystems if needed to demonstrate full functionality of the subsystem.	High-fidelity lab demonstration or limited/restricted flight demonstration for a relevant environment. Integration of technology is well defined.
7.	System prototype demonstration in a realistic environment.	Prototype near or at planned operational system. Represents a major step up from technology readiness level 6, requiring the demonstration of an actual system prototype in a realistic environment, such as in an aircraft, vehicle, or space. Examples include testing the prototype in a test bed aircraft.	Prototype. Should be form, fit, and function integrated with other key supporting elements/subsystems to demonstrate full functionality of subsystem.	Flight demonstration in representative realistic environment such as flying test bed or demonstrator aircraft. Technology is well substantiated with test data.
8.	Actual system completed and "flight qualified" through test and demonstration.	Technology has been proven to work in its final form and under expected conditions. In almost all cases, this technology readiness level represents the end of true system development. Examples include developmental test and evaluation of the system in its intended weapon system to determine if it meets design specifications.	Flight qualified hardware.	Developmental Test and Evaluation in the actual system application.
9.	Actual system "flight -proven" through successful mission operations.	Actual application of the technology in its final form and under mission conditions, such as those encountered in operational test and evaluation. In almost all cases, this is the end of the last "bug fixing" aspects of true system development. Examples include using the system under operational mission conditions.	Actual system in final form.	Technology assessed as fully mature. Operational Test and Evaluation in operational mission conditions.

Source: GAO analysis of National Aeronautics and Space Administration data. | GAO-18-280SP.

Appendix V: Elements of a Sound Business Case

The development and execution of a knowledge-based business case for the National Aeronautics and Space Administration's (NASA) projects can provide early recognition of challenges, allow managers to take corrective action, and place needed and justifiable projects in a better position to succeed. Our prior work of best practice organizations shows the risks inherent in NASA's work can be mitigated by developing a solid, executable business case before committing resources to a new product's development.[53] In its simplest form, a knowledge-based business case is evidence that (1) the customer's needs are valid and can best be met with the chosen concept and that (2) the chosen concept can be developed and produced within existing resources—that is, proven technologies, design knowledge, adequate funding, adequate time, and adequate workforce to deliver the product when needed. A program should not be approved to go forward into product development unless a sound business case can be made. If the business case measures up, the organization commits to the development of the product, including making the financial investment. The building of knowledge consists of information that should be gathered at these three critical points over the course of a program:

- When a project begins development, the customer's needs should match the developer's available resources—mature technologies, time, and funding. An indication of this match is the demonstrated maturity of the technologies required to meet customer needs— referred to as critical technologies. If the project is relying on heritage—or pre-existing—technology, that technology must be in the appropriate form, fit, and function to address the customer's needs within available resources. The project will generally enter development after completing the preliminary design review, at which time a business case should be in hand.
- Then, about midway through the project's development, its design should be stable and demonstrate it is capable of meeting performance requirements. The critical design review takes place at that point in time because it generally signifies when the program is ready to start building production-representative prototypes. If project development continues without design stability, costly redesigns to address changes to project requirements and unforeseen challenges can occur.
- Finally, by the time of the production decision, the product must be shown to be producible within cost, schedule, and quality targets and have demonstrated its reliability, and the design must demonstrate that it performs as needed through

---

[53] GAO, *Defense Acquisitions: Key Decisions to be Made on Future Combat System,* GAO-07-376 (Washington, D.C.: Mar. 15, 2007); *Defense Acquisitions: Improved Business Case Key for Future Combat System's Success,* GAO-06-564T (Washington, D.C.: Apr. 4, 2006); *NASA: Implementing a Knowledge-Based Acquisition Framework Could Lead to Better Investment Decisions and Project Outcomes,* GAO-06-218 (Washington, D.C.: Dec. 21, 2005); and *NASA's Space Vision: Business Case for Prometheus 1 Needed to Ensure Requirements Match Available Resources,* GAO-05-242 (Washington, D.C.: Feb. 28, 2005).

realistic system-level testing. Lack of testing increases the possibility that project managers will not have information that could help avoid costly system failures in late stages of development or during system operations.

APPENDIX VI: COMMENTS FROM THE NATIONAL AERONAUTICS AND SPACE ADMINISTRATION

National Aeronautics and Space Administration
Office of the Administrator
Washington, DC 20546-0001

April 13, 2018

Ms. Cristina T. Chaplain
Director
Contracting and National Security Acquisitions
United States Government Accountability Office
Washington, DC 20548

Dear Ms. Chaplain:

The National Aeronautics and Space Administration (NASA) appreciates the opportunity to comment on the Government Accountability Office (GAO) draft report entitled "NASA: Assessments of Major Projects" (GAO-18-280SP). This assessment provides NASA with a valued independent perspective on our major acquisitions. We appreciate the open and constructive dialogue between NASA and the GAO engagement team, and we look forward to continuing to work with the GAO to identify and address any challenges that may enable cost and schedule improvements in our current and future projects.

This year's report represents the tenth annual iteration of the GAO's legislatively mandated assessment of NASA's major acquisitions. The GAO has provided NASA with several highly valued insights into various aspects of our acquisition approaches since the inaugural report's issuance in 2009, many of which have resulted in programmatic improvements and enhancements. However, the number of projects in this year's assessment portfolio is 36 percent higher than the average for this engagement. This growth may be attributed to a greater number of major spaceflight development projects in NASA's portfolio, but another driver has been the expansion of the legislative mandate to include potential major projects in formulation; technology demonstrations; and communications, tracking, and infrastructure refreshment activities. We are pleased the GAO makes a dedicated effort to minimize disruption to projects wherever possible; however, there are times where disruption is unavoidable, particularly with regard to projects earlier in their life cycles prior to confirmation. While NASA appreciates the clear demonstrated value in this recurring engagement, it is equally important to recognize the growing administrative burden that is necessary to respond to GAO inquiries with valid, accurate, and timely information.

NASA appreciates the GAO's recognition of alternative approaches to managing different types of projects, such as technology demonstration spacecraft and experimental aircraft. The GAO has established best practices regarding the maturation of technologies to Technology Readiness Level 6 (TRL-6) by the Preliminary Design Review (PDR) for science and human exploration missions, and we strive to achieve this technology maturation schedule on our spaceflight missions per the NASA Procedural Requirement

(NPR) 7120.5E policy. As the GAO has now incorporated technology demonstration projects into the audit, it is important to remember that the purpose of these activities is to mature specific new technologies to TRL-6 or higher by the end of their demonstration phase in a relevant environment. As such, at PDR for these technology demonstration projects, it is anticipated that technologies would not necessarily be at TRL-6 unlike for science and human exploration missions. NASA looks forward to working with the GAO to ensure a common understanding of the ways our approach to these types of projects differ from the typical NPR 7120.5E approach.

NASA is in agreement with the GAO that there are several large, complex projects in development that comprise a large relative share of the total current major project portfolio. We also recognize some development challenges exist in these projects that have resulted in recently revised cost and schedule plans in some instances. As we work to mitigate cost, schedule, and technical risk in these critical development efforts, we would like to take this opportunity to highlight the overall positive trending of our cost and schedule estimation practices over the past twenty years. When taken in aggregate, NASA's cost and schedule performance has markedly improved, moving from 45 percent average development cost growth against established baselines in the pre-cost confidence level era in the early 2000s, to less than 2 percent average development growth since the advent of the Joint Cost and Schedule Confidence Level (JCL) policy implementation in 2009. We are continuing to assess and advance the Agency's programmatic capabilities across the cost-estimating, pricing, scheduling, and earned value management communities, as well as any strategic workforce implications or augmentations required to support future mission needs. NASA looks forward to continued maturation and application of our cost and schedule estimation practices and tools as additional large acquisitions continue along their life cycles in the years ahead.

The majority of work on NASA's new deep space exploration system is on track. As a percentage of overall effort, we are seeing specific areas of targeted challenges consistent with a first-time design, development, test, and build of a human spacecraft system for deep space, which offer future opportunity for subsequent builds. NASA is using lessons learned from these experiences to drive efficiencies into overall production and operations planning. With NASA's multi-mission approach to deep space exploration, we have hardware in production for the first and second missions, and we are gearing up for the third flight. As teams complete hardware for one flight, they move on to the next set of flights. Extreme focus on early flights can be harmful to future missions in a multi-flight program, and focusing solely on the first flights of Exploration Mission 1 (EM-1) for the Space Launch System (SLS) and EM-2 for Orion can impact the ability to deliver a human deep space exploration system for use in the decades to come. Taking the time to effectively resolve first-time build challenges up-front leads to short-term schedule and cost challenges but yields significant benefits for out-year flights.

After a thorough schedule analysis in 2017, NASA updated the EM-1 launch date and provided the required Congressional notification per the 2005 NASA Authorization Act, Section 103. The notification included updated cost and schedule estimates for the SLS and Exploration Ground Systems (EGS) programs through EM-1 because these programs are baselined to the EM-1 launch date. An updated cost estimate for Orion was not included in the initial notification because this program is baselined through EM-2, a different development duration than the SLS EM-1 baseline. As part of NASA's normal budgeting process, Orion is refining its EM-2 baseline cost estimate to incorporate the new schedule planning for EM-1 and account for post-EM-1 activities. NASA will provide the results of this update when it is complete.

NASA would like to thank the GAO for continuing to work with project subject-matter experts to consider and incorporate technical corrections as part of this audit. We appreciate the consideration of these comments, which is important for an accurate and balanced presentation of the projects' technical status. We look forward to working with the GAO to ensure the technical review process continues to add value in the future.

NASA greatly appreciates the ongoing dialogue with the GAO on this critical engagement and is committed to working jointly to address any questions or concerns related to this effort. Please contact Kevin M. Gilligan at (202) 358-4544 if you have any questions or require additional information.

Sincerely,

Stephen G. Jurczyk
Associate Administrator (Acting)

In: NASA: Major Projects and Select Issues
Editor: Oliver Cummings

ISBN: 978-1-53614-817-6

Chapter 2

NASA MAJOR PROJECTS: PORTFOLIO IS AT RISK FOR CONTINUED COST GROWTH AND SCHEDULE DELAYS*

Cristina T. Chaplain

WHY GAO DID THIS STUDY

GAO designated NASA's acquisition management as a high-risk area in 1990 after a history of persistent cost growth and schedule slippage in many of NASA's major projects. In more recent years, GAO found that NASA had taken some steps to improve its management, and, in May 2017, GAO found that projects were continuing a generally positive trend of limiting cost and schedule growth. But at the same time, GAO noted that many of these projects, including some of the most expensive ones, were approaching the phase in their life cycles when cost and schedule growth is most likely.

This chapter summarizes GAO's 2018 findings from its 10th annual snapshot of how well NASA is planning and executing its major acquisition projects, and describes (1) the cost and schedule performance of NASA's portfolio of major projects and (2) the extent to which NASA faces risks for further cost increases and schedule delays. To conduct its review for the 2018 report, GAO-18-280SP, GAO analyzed cost, schedule, and other data for NASA's 26 major projects, each with a life-cycle cost of over $250 million; reviewed monthly project status reports; and interviewed NASA officials.

* This is an edited, reformatted and augmented version of United States Government Accountability Office Testimony Before the Subcommittee on Space, Committee on Science, Space, and Technology, House of Representatives, Publication No. GAO-18-576T, dated Thursday, June 14, 2018.

What GAO Recommends

GAO is not making any new recommendations in this statement. GAO has made recommendations in prior reports to strengthen NASA's acquisition management of its major projects. NASA generally agreed with these recommendations, but has not fully addressed some of them. GAO continues to believe they should be fully addressed.

What GAO Found

The cost and schedule performance of the National Aeronautics and Space Administration's (NASA) portfolio of major projects has deteriorated, but the extent of cost performance deterioration is unknown. NASA expects cost growth for the Orion crew capsule—one of the largest projects in the portfolio—but does not have a current cost estimate. In addition, the average launch delay for the portfolio was 12 months, the highest delay GAO has reported in its 10 years of assessing major NASA projects (see figure below).

The deterioration in portfolio performance was the result of 9 of the 17 projects in development experiencing cost or schedule growth.

NASA's Major Project Portfolio Cost and Schedule Performance Deteriorated in 2018

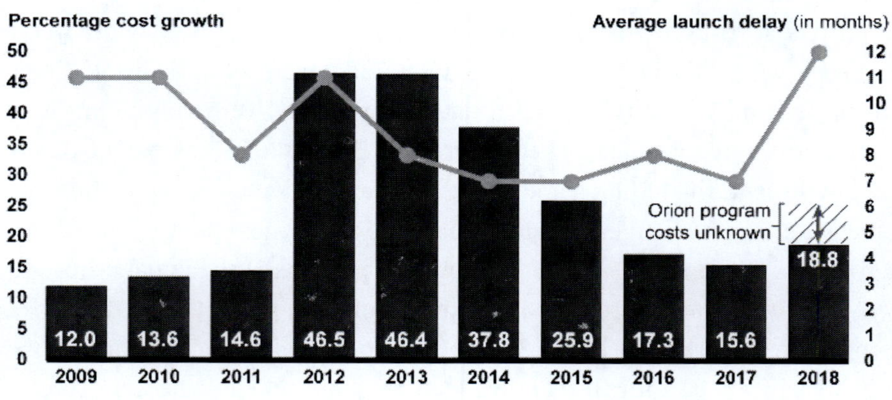

Source: GAO analysis of National Aeronautics and Space Administration (NASA) data. | GAO-18-576T.

- Four projects encountered technical issues that were compounded by risky program management decisions. For example, the Space Launch System and Exploration Ground Systems programs are large-scale, technically complex human spaceflight programs, and NASA managed them to aggressive schedules and with insufficient levels of cost and schedule reserves. This made it more difficult for the programs to operate within their committed baseline cost and schedule estimates.

- Two projects ran into technical challenges that resulted in delays in the integration and test phase. For example, in December 2017, GAO found that the James Webb Space Telescope project encountered delays primarily due to the integration of the various spacecraft elements taking longer than expected, as well as the need to resolve technical issues during testing. GAO has previously found that integration and testing is when projects are most at risk of incurring cost and schedule growth.
- Three projects experienced cost growth or schedule delays due to factors outside of the projects' control, such as delays related to their launch vehicles.

NASA continues to face increased risk of cost and schedule growth in future years due to new, large and complex projects that will enter the portfolio and expensive projects remaining in the portfolio longer than expected.

NASA's major projects are the key enablers for the agency to achieve its vision and its mission. They include NASA's Space Launch System (SLS) and Orion Multi-Purpose Crew Vehicle (Orion), which are the centerpieces of NASA's human exploration plans; Mars 2020 and Europa Clipper, which will further our understanding of the habitability of other planets; and the Ice, Cloud, and Land Elevation Satellite-2 (ICESat-2), which will provide better data on changes in the Earth. In its fiscal year 2019 budget request, NASA requested $19.9 billion, which included about $4.5 billion for its deep space exploration programs and research and about $6 billion for science programs and research. As these projects are complex and specialized, and often push the state of the art in space technology, NASA manages a portfolio that will always have inherent technical, design, and integration risks.

We have been assessing the cost and schedule performance of NASA's major projects—those that have a life-cycle cost over $250 million— annually for 10 years. Over this time, we have seen NASA make progress in reducing acquisition risks, such as by improving cost and schedule estimating tools, and establishing design metrics and tracking projects against the metrics. But our most recent assessment in May 2018 found that the cost and schedule performance of the portfolio deteriorated after several years of following a generally positive trend of limiting cost and schedule growth.[1]

My statement today is based primarily upon our May 2018 report. Specifically, I will discuss (1) the cost and schedule performance of NASA's portfolio of major projects and (2) the extent to which NASA faces risks for continued cost increases and schedule delays. To conduct this work, we collected cost and schedule information from NASA's major projects using a data collection instrument, analyzed projects' monthly status reports,

[1] GAO, *NASA: Assessments of Major Projects*, GAO-18-280SP (Washington, D.C.: May 1, 2018).

interviewed NASA project and headquarters officials, and reviewed project documentation. At the time of our review, there were 26 major projects in total, but the information available depended on where a project was in its life cycle.[2] For the 17 projects that were in the implementation phase, we compared current cost and schedule estimates to their original cost and schedule baselines. We reviewed historical data on cost and schedule performance for major projects from our prior reports and compared them to the performance of NASA's current portfolio of major projects. More detailed information on our objectives, scope, and methodology for that work can be found in our May 2018 report. In addition, we requested and received an update from NASA in early June 2018 on whether the agency had completed a new life-cycle cost estimate for the Orion program. We also updated the actual launch date for the Gravity Recovery and Climate Experiment Follow-On (GRACE-FO) project and updated information on the launch date for the Ionospheric Connection Explorer (ICON) project.

We conducted the work on which this statement is based in accordance with generally accepted government auditing standards. Those standards require that we plan and perform the audit to obtain sufficient, appropriate evidence to provide a reasonable basis for our findings and conclusions based on our audit objectives. We believe that the evidence obtained provides a reasonable basis for our findings and conclusions based on our audit objectives.

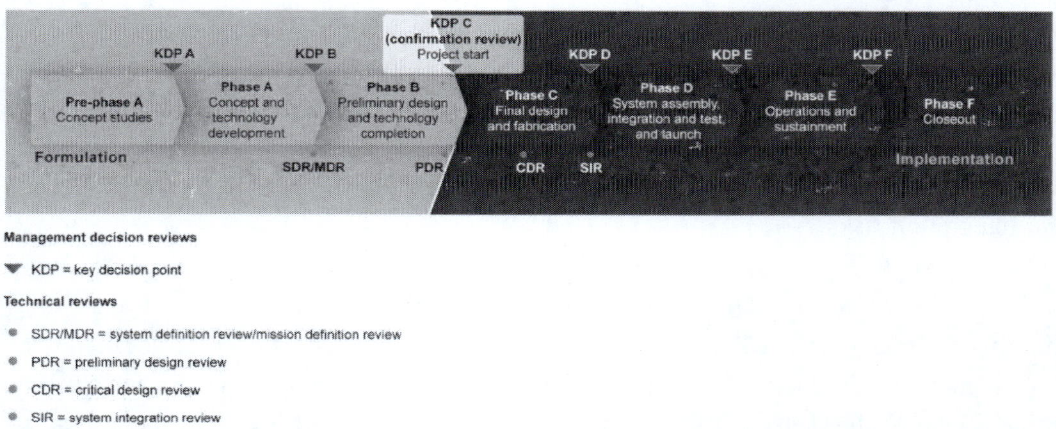

Source: GAO analysis of National Aeronautics and Space Administration (NASA) data. | GAO-18-576T.

Figure 1. NASA's Life Cycle for Space Flight Projects.

[2] Eight projects were in an early stage of development, called formulation, when there are still unknowns about requirements, technology, and design. For those projects, we reported preliminary cost ranges and schedule estimates. The Commercial Crew Program has a tailored project life cycle and project management requirements. As a result, it was excluded from our cost and schedule performance analysis.

BACKGROUND

NASA Acquisition Life Cycle for Space Flight Projects

The life cycle for NASA space flight projects consists of two phases— formulation, which takes a project from concept to preliminary design, and implementation, which includes building, launching, and operating the system, among other activities. NASA further divides formulation and implementation into phases, phase A through phase F. Major projects must get approval from senior NASA officials at key decision points before they can enter each new phase. Formulation culminates in a review at key decision point C, known as project confirmation, where cost and schedule baselines are established and documented in a decision memorandum. Figure 1 depicts NASA's life cycle for space flight projects.

At the time of our review in May 2018, NASA had a portfolio of 26 major projects (see table 1). See appendix I for a brief description of each project.

NASA Acquisition Management as a High-Risk Area

NASA acquisition management is an area that we monitor on our high-risk list.[3] Our high-risk series is a biennial report that keeps focused attention on government operations with greater vulnerabilities to fraud, waste, abuse, and mismanagement or that are in need of transformation to address economy, efficiency, or effectiveness challenges. In 1990, we first designated the area as high risk because there was little emphasis on end results, product performance, and cost control; the acquisition process itself was cumbersome and time-consuming; and NASA found itself procuring expensive hardware that did not work properly. For example, in April 1990, NASA deployed the $1.5 billion Hubble Space Telescope and soon after, the agency discovered that the primary mirror had been manufactured in the wrong shape, severely degrading some of the telescope's scientific capabilities.

Subsequently, we and other organizations, including the National Academy of Sciences and NASA's Office of the Inspector General, found that NASA's cost estimates were overly optimistic.[4] Our reviews also found that NASA continued to experience

---

[3] GAO, *High-Risk Series: Progress on Many High-Risk Areas, While Substantial Efforts Needed on Others*, GAO-17-317 (Washington, D.C.: Feb. 15, 2017). We first designated NASA contract management as a high-risk area in 1990. In 2009, we updated the title of the area to NASA acquisition management because of the scope of issues that needed to be resolved to address persistent cost growth and schedule delays.

[4] GAO, Space Programs: NASA's Independent Cost Estimating Capability Needs Improvement, GAO/NSIAD-93-73 (Washington, D.C.: Nov. 5, 1992); NASA Program Costs: Space Missions Require Substantially more Funding Than Initially Estimated, GAO/NSIAD-93-97 (Washington, D.C.: Dec. 31, 1992); National Research Council of the National Academies, Controlling Cost Growth of NASA Earth and Space Science Missions

significant cost and schedule growth due, in part, to not having a disciplined cost estimating process.

- In 1992, we reviewed the cost and schedule performance of 29 NASA programs and found that 25 of those programs experienced cost growth that ranged from 14 to 426 percent above their initial estimates.[5] Further, the median estimate change for all programs was an increase of 77 percent. General reasons that NASA provided for the cost growth included insufficient definition studies, program and funding instability, overly optimistic assumptions by program officials, and unrealistic contractor estimates. The more specific reasons for the cost growth we found included program redesigns, technical complexities, budget constraints, and incomplete cost estimates.
- In 2004, we reviewed the cost and schedule performance of 27 NASA programs and found that 17 of the programs experienced cost growth.[6] Cost growth for 10 of the 17 programs was over 25 percent. We found that considerable change in NASA's program cost estimates—both increases and decreases—indicated that NASA lacked a clear understanding of how much its programs cost and how long they will take to achieve their objectives. Further, we found that NASA's basic cost-estimating processes—an important tool for managing programs—lacked the discipline needed to ensure that program estimates are reasonable.

In more recent years we have found that NASA's leadership was focused on improving acquisition outcomes and had taken some steps to improve its management.

- In 2006, NASA established a management review process to enable NASA's senior management to more effectively monitor a project's performance, including cost, schedule, and cross-cutting technical and nontechnical issues.
- In 2009, NASA began requiring that NASA major programs and projects develop a joint cost and schedule confidence level (JCL) prior to project confirmation in order to ensure that cost and schedule estimates were realistic and projects thoroughly planned for anticipated risks. The JCL is a point-in-time estimate that, among other things, includes all cost and schedule elements, incorporates and quantifies known risks, assesses the impacts of cost and schedule to date, and addresses available annual resources. NASA policy generally requires that projects be baselined and budgeted at the 70 percent confidence level.

(Washington, D.C.: National Academies Press, 2010); and NASA Office of Inspector General, Inspector General Assessment of NASA's Most Serious Management and Performance Challenges, (Washington, D.C.: Feb. 4, 2002) and NASA's Independent Cost Estimating Capability, IG-00-045 (Washington, D.C.: Sept. 20, 2000).

[5] GAO/NSIAD-93-97.

[6] GAO, NASA: Lack of Disciplined Cost-Estimating Processes Hinders Effective Program Management, GAO-04-642 (Washington, D.C.: May 28, 2004).

- In 2012, the agency established metrics to more consistently measure a project's design progress and, in 2014, we found that most major projects in the portfolio were tracking and reporting those metrics. In addition, experts with whom we met confirmed that NASA's metrics are valid measures to assess design maturity in space systems.
- Since 2015, we have observed a positive trend of higher numbers of projects maturing technologies prior to preliminary design review.[7] Demonstrating that technologies will work as intended in a relevant environment serves as a fundamental element of a sound business case, and projects falling short of this standard often experience subsequent technical problems. Our best practices work has shown that maturing technologies prior to preliminary design review can minimize risks for projects entering development, which lowers the risk of subsequent cost growth and schedule delays.

Table 1. Major NASA Projects Reviewed in GAO's 2018 Assessment by Phase

Projects in formulation	Double Asteroid Redirection Test (DART) Europa Clipper Low Boom Flight Demonstrator (LBFD) Lucy Plankton, Aerosol, Cloud, ocean Ecosystem (PACE) Psyche Restore-L Wide-Field Infrared Survey Telescope (WFIRST)
Projects in implementation	Commercial Crew Program (CCP) Exploration Ground Systems (EGS) Gravity Recovery and Climate Experiment Follow-On (GRACE-FO) Ice, Cloud, and Land Elevation Satellite-2 (ICESat-2) Interior Exploration using Seismic Investigations, Geodesy, and Heat Transport (InSight) Ionospheric Connection Explorer (ICON) James Webb Space Telescope (JWST) Landsat 9 (L9) Laser Communications Relay Demonstration (LCRD) Mars 2020 NASA ISRO Synthetic Aperture Radar (NISAR) Orion Multi-Purpose Crew Vehicle (Orion) Parker Solar Probe (PSP) (formerly Solar Probe Plus) Radiation Budget Instrument (RBI) Space Launch System (SLS) Space Network Ground Segment Sustainment (SGSS) Surface Water and Ocean Topography (SWOT) Transiting Exoplanet Survey Satellite (TESS)

Source: GAO analysis of National Aeronautics and Space Administration (NASA) data. | GAO-18-576T.

[7] A technology is considered mature when it reaches a technology readiness level 6, which is achieved after demonstrating a representative prototype of the technology in a relevant environment that simulates the harsh conditions of space.

We believe that many of these steps NASA has taken contributed to the largely positive trend of cost and schedule performance for NASA's portfolio of major projects between 2013 and 2017. In our May 2017 assessment of major projects, we found that out of 16 projects in development, 5 experienced cost growth and 4 experienced schedule delays over their development cost and schedule baselines.[8] Both of these measures were at or near the lowest levels we have reported since we began our annual assessments in 2009.

However, we also found in our February 2017 high risk update that NASA needed to do more with respect to anticipating and mitigating risks— especially with regard to large programs, estimating and forecasting costs for its largest projects, and implementing management tools.[9] We highlighted several actions that would be critical to improving NASA's acquisition outcomes, including the following:

- Ensuring that NASA conducted adequate and ongoing assessments of risks for larger programs because the impacts of any potential miscalculations will be felt across NASA's portfolio.
- Ensuring that NASA understood long-term human exploration program costs. While the three major human exploration programs— Orion, SLS, and the Exploration Ground Systems (EGS)—have been baselined, none of the three programs has a baseline that covers activities beyond the second planned flight. Long-term estimates, which could be revised as potential mission paths are narrowed and selected, would provide decision makers with a more informed understanding of costs and schedules associated with potential agency development paths.
- Ensuring that program offices regularly and consistently updated their JCL across the portfolio. As a project reaches the later stages of development, especially integration and testing, its risk posture may change. An updated project JCL would provide both project and agency management with data on relevant risks that can guide project decisions.
- Ensuring that NASA continued its efforts to build capacity in areas such as cost and schedule estimating and measuring contractor performance.

Further, in our 2016 and 2017 assessments of major projects, we found that while the cost and schedule performance of NASA's portfolio was improving, a number of large, complex projects were in or would soon be entering the integration and test phase—the phase in development that often reveals unforeseen challenges that can lead to cost and schedule growth.[10] In May 2017, projects in this phase included all three human spaceflight

---

[8] GAO, *NASA: Assessments of Major Projects*, GAO-17-303SP (Washington, D.C.: May 16, 2017).

[9] GAO-17-317.

[10] GAO, *NASA: Assessments of Major Projects,* GAO-16-309SP (Washington, D.C.: Mar. 30, 2016); and GAO-17-303SP.

programs and the James Webb Space Telescope (JWST).[11] Subsequently, we found that these programs experienced delays during this phase of development. For example, in December 2017, NASA announced a 13- to 19-month delay for the first integrated mission of Orion, SLS, and EGS. This mission is referred to as Exploration Mission 1 (EM-1) and will not have crew. In addition, in December 2017, we found that the JWST project continued to make progress towards launch, but the program was encountering technical challenges that required both time and money to fix and may lead to additional delays.[12] Subsequently, the JWST project delayed its launch readiness date by at least 19 months from October 2018 to May 2020.

PORTFOLIO COST AND SCHEDULE DETERIORATED BUT EXTENT OF COST GROWTH IS UNKNOWN

The cost and schedule performance of NASA's portfolio of major projects deteriorated between May 2017 and May 2018, but the extent of cost growth is unknown. NASA lacks a current cost estimate for its Orion crew capsule—one of the largest programs in the portfolio—but expects the program will exceed its cost baseline when NASA updates the program's life-cycle cost estimate. Because the Orion program accounts for about 22 percent of all development costs, even a small percentage of cost growth for the Orion program could significantly affect portfolio cost performance. The known negative cost and schedule performance is largely driven by the cost and schedule growth of four projects—SLS, EGS, Space Network Ground Segment Sustainment (SGSS) and Mars 2020—that experienced technical problems compounded by programmatic challenges. Together, these projects experienced $638 million in cost growth and 59 months in aggregate schedule delays. Two projects—JWST and ICESat-2—experienced schedule delays due to technical challenges identified during integration and test. Another 3 projects—NASA Indian Space Research Organisation Synthetic Aperture Radar (NISAR), ICON, and GRACE-FO—experienced cost growth or delays largely due to factors outside of the projects' control, such as launch vehicle delays.

---

[11] GAO-17-303SP.

[12] GAO, NASA: Preliminary Observations on the Management of Space Telescopes, GAO-18-277T (Washington, D.C.: Dec. 6, 2017).

Portfolio Average Launch Delays Increased, but NASA Lacks a Current Orion Program Cost Estimate to Determine Extent of Cost Growth

The average launch delay increased from 7 months in our May 2017 report to 12 months in our May 2018 report—the highest schedule delay we have reported to date.[13] We were not able to determine the extent of portfolio cost growth this year because NASA does not have a current cost estimate for the Orion program—one of the largest programs in its portfolio—and officials expect the cost to increase. As of June 2017, the Orion program's development cost was about $6.6 billion; based on that estimate, it accounts for 22 percent of the portfolio's estimated $30.1 billion of development costs. As a result, a small percentage of cost growth for the Orion program could significantly affect cost performance. Even without including Orion cost growth, the overall development cost growth for the portfolio of 17 development projects increased to 18.8 percent, up from 15.6 percent in 2017 (see figure 2).[14]

Senior-level NASA officials told us they expect that the Human Exploration and Operations Mission Directorate and the Orion program will complete an updated life-cycle cost estimate in June 2018. This would be approximately 10 months after the program raised to senior-level officials' attention that the program expects cost growth over its cost baseline during an August 2017 briefing concerning potential cost increases related to the launch delay for EM-1.[15] In early June 2018, NASA officials said that they had not yet completed the updated life-cycle cost estimate.

In our May 2018 report, we found that 7 of 17 NASA major projects had stayed within cost and schedule estimates since our 2017 annual assessment of major projects, but 9 projects experienced cost growth or schedule delays and cost growth is expected for the Orion program. Table 2 provides data on the cost and schedule performance between our May 2017 and 2018 reports for the 17 major projects in development that have cost and schedule baselines.[16]

---

[13] GAO-17-303SP.

[14] We have historically presented cost and schedule performance including and excluding the James Webb Space Telescope (JWST) because, prior to 2015, it had a development cost baseline significantly larger than other projects and the magnitude of its cost growth masked the performance of the remainder of the portfolio. Now that other projects in the portfolio, such as Orion and the Space Launch System, have large development cost baselines, we no longer present cost performance trends excluding JWST.

[15] The Orion program is baselined to the second combined mission of Orion, SLS, and EGS. This mission is known as Exploration Mission 2 (EM-2) and will have crew.

[16] GAO-17-303SP.

Table 2. Development Cost and Schedule Performance of Selected Major NASA Projects in Development as of May 2018 Assessment

Overall performance	Project	Confirmation date	Changes between May 2017 and May 2018		Cumulative performance	
		Year	Cost (millions)	Schedule (months)	Cost (millions)	Schedule (months)
Lower than expected cost	PSP	2014	$0.0	0	-$5.4	0
	GRACE-FO	2014	-$1.6	3	-$2.2	3
	TESS	2014	-$13.1	-2	-$39.9	-2
Within baseline	ICON	2014	$0.0	8	$0.0	8
	SWOT	2016	$0.0	0	$0.0	0
	LCRD	2017	$0.0	0	$0.0	0
	Landsat 9	2017	$0.0	0	$0.0	0
Higher than expected cost	Mars 2020[a]	2016	$12.9	0	$10.7	0
	NISAR	2016	$22.0	0	$22.0	0
Replan[c]	InSight	2014	$0.0	0	$131.7	26
	EGS (EM-1)	2014	$417.8	19	$421.4	19
	SLS (EM-1)	2014	$147.8	19	$147.8	19
Rebaseline[c]	JWST	2008	$0.0	19	$3,607.7	71
	ICESat-2	2012	$1.4	4	$206.3	17
	SGSS[b]	2013	$59.5	21	$421.6	48
Canceled	RBI	2016	$0.0	0	$0.0	0
Under revision	Orion (EM-2)[d]	2015	$0.0	0	-$151.7	0
Total:			$646.7	91	$4,770.0	209

Legend: PSP: Parker Solar Probe; GRACE-FO: Gravity Recovery and Climate Experiment Follow-On; TESS: Transiting Exoplanet Survey Satellite; ICON: Ionospheric Connection Explorer; SWOT: Surface Water and Ocean Topography; NISAR: NASA Indian Space Research Organisation – Synthetic Aperture Radar; LCRD: Laser Communications Relay Demonstration; InSight: Interior Exploration using Seismic Investigations, Geodesy, and Heat Transport; EGS: Exploration Ground Systems; SLS: Space Launch System; EM-1: Exploration Mission 1; JWST: James Webb Space Telescope; ICESat-2: Ice, Cloud, and Land Elevation Satellite-2; SGSS: Space Network Ground Segment Sustainment; RBI: Radiation Budget Instrument; Orion: Orion Multi-Purpose Crew Vehicle; EM-2: Exploration Mission 2.

Source: GAO analysis of National Aeronautics and Space Administration (NASA) data. | GAO-18-576T.

Note: The confirmation date is the year NASA established and documented a cost and schedule baseline for each project. Positive values indicate cost growth or launch delays. Negative values indicate cost decreases or earlier than planned launch dates.

[a] The Mars 2020 project used $2.2 million in funds originally budgeted for development for formulation activities. This partially offsets an increase of $12.9 million in development cost growth primarily due to increased costs associated with a technology demonstration instrument and entry, descent, and landing instrument.

[b] The SGSS project reported cost growth up through its first operational readiness review, which is currently planned for the end of fiscal year 2019. However, the project expects that there could be additional cost and schedule growth beyond what is reported here.

[c] A replan process is initiated if development costs increase by 15 percent or more. NASA replanned the SLS program even though development costs did not increase by 15 percent or more. A replan does not require a new project baseline to be established. A rebaseline is a process initiated if development costs increase by 30 percent or more. When development cost growth is likely to exceed the development cost estimate by 15 percent or more, NASA must submit a report to the Committee on Science, Space, and Technology of the House of Representatives and the Committee on Commerce, Science, and Transportation of the Senate. In addition, if a project or program milestone is likely to be delayed by 6 months or more, this report is also required.

[d] NASA officials said they are revising the Orion program's life-cycle cost estimate and expect to complete a new estimate in June 2018. The new cost is expected to exceed the program's development cost baseline. The current costs in the table reflect the estimate provided in June 2017. The cumulative cost change reflects the program shifting $151.7 million of funding previously budgeted for the development phase to the formulation phase.

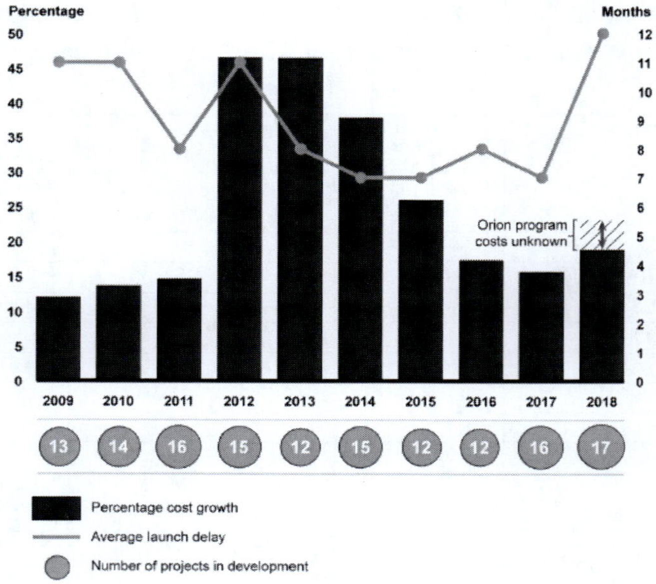

Source: GAO analysis of National Aeronautics and Space Administration (NASA) data. | GAO-18-576T.

Note: The years given in the figure refer to the year we issued each of our annual assessments of NASA major projects reports. Cost and schedule performance is compared across each report period.

Figure 2. Development Cost Performance and Average Launch Delay for Major NASA Projects from 2009 to 2018.

The deteriorating cost and schedule performance of the portfolio in 2018 is the result of

- four projects—SLS, EGS, SGSS, and Mars 2020—addressing technical challenges that were compounded by risky programmatic decisions;
- two projects—JWST and ICESat-2—experiencing delays due to technical challenges identified during integration and test; and
- three projects—NISAR, ICON, and GRACE-FO—experiencing cost growth or delays largely due to factors outside of the projects' control.

We elaborate on these three scenarios above.

Technical Challenges Compounded by Risky Programmatic Decisions

Together, SLS, EGS, SGSS, and Mars 2020 experienced $638 million in cost growth and 59 months in aggregate schedule delays due to technical problems that were compounded by programmatic challenges since our May 2017 report.[17] The SLS and EGS programs experienced cost growth and schedule delays associated with EM-1, their first combined mission along with the Orion program. We have found for several years that the

[17] GAO-17-303SP.

human spaceflight programs—Orion, SLS, and EGS—are making progress maturing designs and building hardware, but also are experiencing some significant engineering and manufacturing challenges. For example, the SLS program ran into numerous challenges completing the welding of its core stage element in 2017. The program stopped welding on the core stage for months to identify and resolve low weld strength in the liquid oxygen and liquid hydrogen tanks due to low weld strength measurements found in the liquid oxygen tanks caused by a program and contractor decision to change the weld tool configuration during fabrication. The EGS program also experienced technical challenges, including with the design and installation of the ground support equipment and the 10 umbilicals that connect SLS and Orion to the Mobile Launcher—which supports the assembly, testing, and servicing of SLS and provides the platform on which SLS and Orion will launch.

Finally, although the Orion program has not yet reported cost growth, it also experienced technical challenges. These challenges included software and hardware delays, and at least 14 months of delays with the European Service Module—which provides air, water, power, and propulsion to Orion during in-space flight—since the element's critical design review in June 2016. In April 2017, we found that, according to program officials, the delays with the service module were largely due to NASA, the European Space Agency, and the European Space Agency contractor underestimating the time and effort necessary to address design issues for the first production service module and the availability of parts from suppliers and subcontractors. NASA expects the Orion program to experience cost growth over its cost baseline to the second combined mission, Exploration Mission 2 (EM-2). However, the extent of the growth is unknown because, as noted above, NASA is currently revising the program's life-cycle cost estimate.

Technical challenges such as these are not unusual for large-scale programs, especially human exploration programs that are inherently complex and difficult. However, we have found that NASA has made programmatic decisions—including establishing low cost and schedule reserves, managing to aggressive schedules, and not following best practices for earned value management or creating reliable cost and schedule baselines—that have compounded the technical challenges (see table 3).[18] As a result, the three human spaceflight programs have been at risk of cost and schedule growth since NASA approved their baselines.

---

[18] GAO, Space Launch System: Management Tools Should Better Track to Cost and Schedule Commitments to Adequately Monitor Increasing Risk, GAO-15-596 (Washington, D.C.: July 16, 2015); NASA Human Space Exploration: Opportunity Nears to Reassess Launch Vehicle and Ground Systems Cost and Schedule, GAO-16-612 (Washington, D.C.: July 27, 2016); Orion Multi-Purpose Crew Vehicle: Action Needed to Improve Visibility into Cost, Schedule, and Capacity to Resolve Technical Challenges, GAO-16-620 (Washington, D.C.: July 27, 2016); and NASA Human Space Exploration: Delay Likely for First Exploration Mission, GAO-17-414 (Washington, D.C.: Apr. 27, 2017). Cost reserves are for costs that are expected to be incurred—for instance, to address project risks—but are not yet allocated to a specific part of the project. Schedule reserves are extra time in project schedules that can be allocated to specific activities, elements, and major subsystems to mitigate delays or address unforeseen risks.

Table 3. Examples of Risky Programmatic Decisions Made by National Aeronautics and Space Administration's (NASA) Human Spaceflight Programs

Programmatic decision	Example(s)	Negative effect	GAO report(s)
NASA baselined the Exploration Ground Systems (EGS), Orion Multi-Purpose Crew Vehicle (Orion), and Space Launch System (SLS) programs with low cost and schedule reserves.	In July 2016, we found that NASA baselined the SLS program with cost reserves of less than 2 percent, even though guidance for Marshall Space Flight Center—the NASA center with responsibility for the SLS program—established standard cost reserve for launch vehicle programs of 20 percent when the baseline is approved.	Operating with low cost and schedule reserves limits a program's ability to address risks and unforeseen technical challenges.	GAO-17-414 and GAO-16-612
NASA managed the EGS, Orion, and SLS programs to an internal schedule for completing development production that was aggressive and could exacerbate delays and lead to cost overruns.	In July 2016, we found that the EGS program planned to conduct the mobile launcher's verification and validation concurrent with ground support equipment systems and umbilicals installation to support the program's internal schedule goal. We found this to be a risky practice because of uncertainties regarding how systems not yet installed may affect the systems already installed.	Working towards a more aggressive internal goal is not a bad practice; however, increasing cost and schedule risk to the program in order to pursue such a goal is not beneficial to programs in the long term.	GAO-16-620 and GAO-16-612
The SLS program did not follow best practices for using earned value management, which integrates the project scope of work with cost, schedule, and performance elements for optimum project planning and control.	In July 2016, we found that the SLS program had not positioned itself well to provide accurate assessments of progress with the core stage because it operated for several years without a performance measurement baseline that is necessary to support full earned value management reporting. The use of earned value management is advocated by both GAO's best practices for cost estimating and NASA's own guidance.	Programs that do not use earned value data are limited in their ability to have accurate assessments of project progress, produce early warning signs of impending schedule delays and cost overruns, and provide unbiased estimates of anticipated costs at completion.	GAO-16-612

Programmatic decision	Example(s)	Negative effect	GAO report(s)
The Orion and SLS programs established baselines that were not fully reliable.[a]	In July 2016, we found that the Orion program did not generally follow best practices in preparing its cost and schedule estimates, which were key inputs into the program's joint cost and schedule confidence level processes and baseline. In July 2015, we found that cost and schedule estimates for the SLS program substantially met five of six characteristics that GAO considers best practices for preparing reliable estimates, but could not be deemed fully reliable because they only partially met the sixth characteristic—credibility.	Without sound cost and schedule estimates, decision makers do not have a clear understanding of the cost and schedule risk inherent in the program or important information needed to make programmatic decisions.	GAO-16-620 and GAO-15-596

Source: GAO analysis of prior GAO reports. | GAO-18-576T.

[a] We did not assess EGS's cost and schedule estimates compared to best practices.

In December 2017, NASA announced the new internal launch readiness date for EM-1 is now December 2019, and has allocated 6 months of schedule reserve available to extend the date to June 2020 for possible manufacturing and production schedule risks. This represents a delay of 13-19 months for EM-1. It is too soon to know if NASA has addressed the programmatic challenges identified above. We will continue to follow up through future reviews.

Similarly, the SGSS project experienced new cost growth of $59.5 million and delayed its completion by 21 months. Project officials attributed the cost growth and delays to the contractor's incomplete understanding of its requirements, which led to poor contractor plans and late design changes. But project management has been a challenge as well.[19] The project has historically struggled to manage contractor performance and has faced both contractor and project staffing shortfalls, as we found in our prior reports starting in 2013.[20] For example, NASA managers noted concerns with contractor plans and staffing estimates in 2013 during project confirmation. In March 2015, we found that the project was being

[19] In 2016, NASA announced it was reclassifying SGSS as a hybrid sustainment project for the Space Network. A hybrid sustainment effort is a sustainment effort that still includes development work. The SGSS project expects to experience additional cost growth and schedule delays, but the exact magnitude is unknown. The project was reevaluating its cost and schedules through its final acceptance review at the time of our review.

[20] GAO, *NASA: Assessments of Selected Large-Scale Projects*, GAO-13-276SP (Washington, D.C.: Apr. 17, 2013); *NASA: Assessments of Selected Large-Scale Projects*, GAO-14-338SP (Washington, D.C.: Apr. 15, 2014); *NASA: Assessments of Selected Large-Scale Projects*, GAO-15-320SP (Washington, D.C.: Mar. 24, 2015); GAO-16-309SP; and GAO-17-303SP.

rebaselined due to the contractor's poor cost and schedule performance and in order to conform with limitations that NASA placed on the funding available to the contractor in fiscal years 2014 and 2015. The contractor was also operating with a limited number of staff at that time. In May 2017, we found that the project continued to experience contractor performance problems and had experienced cost growth and schedule delays over the 2015 rebaseline even as the project decreased its scope. In addition, the project experienced staff shortfalls in key areas, such as systems engineering and business management.

The Mars 2020 project experienced $12.9 million in development cost growth, but no schedule delays. The cost growth was primarily due to technical challenges on a technology demonstration instrument and higher than anticipated integration costs for an entry, descent, and landing instrument. Both instruments are funded by the Human Exploration and Operations and Space Technology Mission Directorates. NASA officials attributed the cost growth of the technology demonstration instrument—which is designed to convert carbon dioxide to oxygen—to the complexity of the technology development for the effort. At the project's preliminary design review in February 2016, a critical technology for the technology demonstration instrument did not meet the recommended level of maturity, which we have found can increase risk for systems entering product development. The project had matured the technology to this recommended level by its critical design review in February 2017. However, as a result of the focus on maturing this particular technology, other components of the instrument fell behind the planned schedule. Project costs for Mars 2020 also increased for an entry, descent, and landing instrument, due, in part, to cost increases for integration and to add additional staff to the instrument team to maintain schedule.

Finally, the Radiation Budget Instrument project would have likely exceeded its cost baseline if NASA had not decided to cancel the project in January 2018. According to NASA's cancellation memorandum, the project was canceled because of continued cost growth, technical issues, and poor contractor performance. In 2017, we found that the project was working to an aggressive schedule, and the prime contractor continued to experience cost overruns even after NASA added a deputy project manager and increased site visits and meetings with the contractor.[21] Subsequently, the project—which was developing an instrument to be hosted on a National Oceanic and Atmospheric Administration satellite— determined that it would not be able to meet its delivery date for integration with the satellite without requiring additional funding in excess of the project's cost baseline if other technical issues arose. In its cancellation memorandum, NASA stated continuing to fund the project from within the Earth Science Division budget would slow other important activities.

Technical challenges identified during integration and test. The JWST and ICESat-2 projects experienced technical challenges during integration and test that delayed their

---

[21] GAO-17-303SP.

schedules. Both projects were previously rebaselined before entering system-level integration and testing, and the current schedule delays are beyond the new schedules that NASA set for the projects in 2011 for JWST and in 2014 for ICESat2.

- The JWST project delayed its launch readiness date by at least 19 months from October 2018 to May 2020. NASA announced two delays for the project since our portfolio-wide review in May 2017.[22] First, as we found in February 2018, the project delayed its launch readiness date by up to 8 months primarily due to the integration of the various spacecraft elements taking longer than expected.[23] Specifically, execution of spacecraft integration and test tasks, due to complexity of work and cautious handling given the sensitivity of flight hardware, was slower than planned. In addition, before the delay, the project used all of its schedule reserves to its prior launch readiness date. This was the result of various contractor workmanship errors, particularly with respect to the spacecraft propulsion systems, as well as the resolution of various technical issues, including a test anomaly on the telescope and sunshield hardware challenges. Second, in March 2018, NASA announced that it had delayed the project's launch readiness date by an additional 11 months to approximately May 2020 and planned to establish an external independent review board to analyze the project's organizational and technical issues to inform a more specific launch time frame.

  The announcement also stated that after a new launch date is established, NASA would provide a new cost estimate that may exceed the $8 billion congressional cost cap that was established in 2011. NASA plans to finalize the project's cost and schedule estimate by the end of June 2018. Because the additional delays were announced while a draft of our May 2018 report was with NASA for comment, we plan to follow up on the reasons for the additional delays and the results of the analysis in a future review.

  In our prior assessments of JWST, we have made recommendations with regard to improving cost and schedule estimating, updating risk assessments, and strengthening management oversight. NASA has generally agreed and taken steps to implement a number of our recommendations. For example, in December 2015, we recommended that the JWST project require contractors to identify, explain, and document anomalies in contractor-delivered monthly earned value management reports.[24] NASA concurred with this recommendation and, in February 2016, directed the contractors to implement the actions stated in the

---

[22] GAO-17-303SP.

[23] GAO, James Webb Space Telescope: Integration and Test Challenges Have Delayed Launch and Threaten to Push Costs Over Cap, GAO-18-273 (Washington, D.C.: Feb. 28, 2018).

[24] GAO, James Webb Space Telescope: Project on Track but May Benefit from Improved Contractor Data to Better Understand Costs, GAO-16-112 (Washington, D.C.: Dec. 17, 2015).

recommendation. However, NASA did not implement some recommendations, which if implemented, may have provided insight into the challenges it now faces. For example, in December 2012, we recommended the JWST project update its JCL.[25] Although NASA concurred with this recommendation, it did not take steps to implement it. An updated JCL may have portended the current schedule delays, which could have been proactively addressed by the project.

- The ICESat-2 project delayed its launch readiness date by 4 months from June to October 2018 due to technical issues with its only instrument, the Advanced Topographic Laser Altimeter System. A key part in the instrument's lasers failed during instrument environmental testing, which delayed the project's system integration review—the start of system-level integration and test. The manufacturer determined the primary cause of the anomaly was a flaw in the design of the mount that ensures a component of the optical module remains in a specific, precise position. The spare flight laser encountered the same problem during earlier testing, which indicated a systemic problem. The project redesigned and repaired the lasers and is proceeding through integration and test.

External factors. External factors—including responding to requests for additional data collection and delays due to launch-vehicle related issues—contributed to cost increases or schedule delays for the NISAR, ICON, and GRACE-FO projects.

- The NISAR project experienced cost growth as the result of an increase in the scope of data collection in response to additional data needs being identified by an interagency working group. The additional data include soil moisture and natural hazard data that would be of value for other federal agencies and the science community. NASA officials said the additional funding for development would be used to upgrade the ground stations so that they can receive the additional data at a higher downlink data rate and volume.
- The ICON project missed its committed launch readiness date because of an accident involving its launch vehicle. In January 2017, two of the Pegasus launch vehicle's three stages were involved in a transport accident. The stages were subsequently returned to the launch vehicle contractor facility for inspection and testing, and no damage was found. The project had been on track to launch early. Subsequently, in September 2017, an anomaly found in testing of the launch vehicle bolt cutter assemblies resulted in additional delays. NASA had planned to launch ICON in mid-June 2018, but recently announced a delay after off-nominal

[25] GAO, James Webb Space Telescope: Actions Needed to Improve Cost Estimate and Oversight of Test and Integration, GAO-13-4 (Washington, D.C.: Dec. 3, 2012).

data was observed from the rocket during transit to the launch site. NASA announced a new launch date would be determined at a later date.

- The GRACE-FO project delayed its launch readiness date from February to May 2018 due to issues with its planned launch vehicle and launch site. The launch vehicle is the responsibility of NASA's partner on the project—German Research Centre for Geosciences (GFZ). GRACE-FO had planned to launch at a Russian launch site. In February 2016, GFZ reported that it was notified by the Russian Federal Space Agency that the Dnepr launch vehicle was no longer available for GRACE-FO. GFZ, in June 2016, arranged to launch the two GRACE-FO spacecraft, along with commercial satellites, on a SpaceX Falcon 9. On May 22, 2018, GRACE-FO launched from Vandenberg Air Force Base in California.

In addition, the Commercial Crew Program also experienced delays, which are not included above because the program does not have a schedule baseline. Since the award of the current Commercial Crew contracts in September 2014, the program, Boeing, SpaceX, and multiple independent review bodies have all identified the contractors' delivery schedules as aggressive. In February 2017, we found that Boeing and SpaceX had determined that neither could meet their original 2017 dates for NASA to certify their systems for human spaceflight.[26] In January 2018, we found that both contractors had notified NASA that final certification dates have slipped again and are now in the first quarter of calendar year 2019.[27] The Commercial Crew Program's schedule analysis indicates that certification may be further delayed to December 2019 for SpaceX and February 2020 for Boeing.

NASA Is Likely to Encounter Additional Cost Growth and Schedule Delays

The composition of the portfolio in the coming years is expected to include large and complex projects, putting NASA at risk of continued cost increases and schedule delays. Specifically, NASA plans to have complex projects enter the development portfolio in the next few years as it holds confirmation reviews and set cost and schedule baselines. This includes the Europa Clipper project and potentially the Wide-Field Infrared Survey Telescope (WFIRST) project. In February 2018, the President's 2019 Budget Request proposed canceling the WFIRST project due to the project's significant costs and higher

---

[26] GAO, NASA Commercial Crew Program: Schedule Pressure Increases as Contractors Delay Key Events, GAO-17-137 (Washington, D.C.: Feb. 16, 2017).

[27] GAO, NASA Commercial Crew Program: Continued Delays Pose Risks for Uninterrupted Access to the International Space Station, GAO-18-317T (Washington, D.C.: Jan. 17, 2018).

priorities in the agency. However, the project may continue if funding is received. Together, preliminary estimates indicate that these two projects could cost as much as $7.8 billion. In addition, NASA expects to begin other large, complex projects like the Lunar Orbital Platform-Gateway— currently being discussed as a space station or outpost in lunar orbit— and a Europa Lander project in the coming years. A December 2017 space policy directive also instructed NASA to return astronauts to the moon for long-term exploration and to pursue human exploration of Mars and the broader solar system.

To its credit, NASA recently took steps to put a process in place to control the costs of two projects while in formulation, which may prove useful if properly executed.

- The Europa Clipper project implemented a process whereby cost growth threats would be offset by descoping instruments in whole or in part. For example, if an instrument exceeds its development cost by 20 percent, the project would propose a descope option to NASA that brings instrument cost below that threshold. NASA had not descoped any instruments as of our May 2018 report.
- The WFIRST project is responding to findings from an independent review that was conducted to ensure the mission's scope and required resources are well understood and executable. The review found that the mission scope is understood, but not aligned with the resources provided and concluded that the mission is not executable without adjustments and/or additional resources. For example, the study team found that NASA's current forecasted funding profile for the WFIRST project would require the project to slow down activities starting in fiscal year 2020, which would result in an increase in development cost and schedule. NASA agreed with the study team's results and directed the project to reduce the cost and complexity of the design in order to maintain costs within the $3.2 billion preliminary cost target.

But even with these efforts, NASA's cost and schedule performance may be further tested in upcoming years as some expensive, complex projects linger in the portfolio longer than expected.

- As previously discussed, the Orion program expects cost growth and faces other schedule and technical risks as it moves through the integration and test phase for EM-1 into at least 2019 and then through 2023 for EM-2. As of August 2017, NASA officials expected that new hardware and addressing development challenges would be the factors contributing to increased cost for the program. For example, there was a cost impact when the program moved from a single-piece, or monolithic, heatshield design to one that employs blocks in order to improve its structural strength. Program officials said they are also assessing schedule delays for EM-2, and noted that the EM-2 launch date depends on the outcome of the EM-1 launch date.

- The SLS and EGS programs continue to face cost, schedule, and technical risks as they move through the integration and test phase into at least 2019. For example, SLS will have to complete a "green run" test which requires multiple first-time efforts. Specifically, the test is the culmination of the development effort and includes the core stage integration with its four main engines, fully fueling with cryogenic hydrogen and oxygen, and then firing all four engines for about 500 seconds. NASA currently has no schedule reserve to its target December 2019 launch readiness date for two key areas in the core stage schedule. First, there is no reserve between the end of core stage production and the delivery of the core stage to the test facility. Second, there is no reserve between the end of the testing and delivery to Kennedy Space Center for final integration and testing prior to launch.
- As previously discussed, the JWST project is at risk of exceeding its congressional cost cap, and faces schedule risks as it completes its remaining integration and test work. These activities have taken considerably longer than planned due to a variety of challenges, including reach and access limitations on the flight hardware. Additionally, the project faces significant work ahead. For example, the project must complete integration of spacecraft element hardware and conduct deployment and environmental tests of the integrated sunshield and spacecraft. Further, it must integrate the telescope element with the spacecraft element to form the JWST observatory, and complete another set of challenging environmental tests on the full integrated observatory. At the same time, the project will need to mitigate dozens of remaining hardware and software risks to acceptable levels and address the project's many potential single point failures to the extent possible.
- The SGSS project expects to experience additional cost growth through the final acceptance review because the full scope of the effort has not been included in the cost. NASA only approved its new cost estimate through the initial operational readiness review, currently planned for September 2019. A project official said NASA headquarters asked the project to determine if there are ways to reduce the cost between the operational readiness review and the final acceptance review. NASA plans to conduct an independent review of the project in mid-2018 to inform a decision on whether to continue the project past the operational readiness review. If NASA decides to continue the project past this review, additional cost growth is expected for SGSS when NASA revisits project costs through future budget cycles.

In closing, NASA continues to make improvements to the acquisition management of its portfolio of major projects. However, the deterioration of the cost and schedule performance of NASA's portfolio this year and the likelihood of additional cost growth and schedule delays demonstrate the need for NASA to continue to take actions to further reduce acquisition risk as we and others have recommended. Continuing to improve cost and schedule estimating tools and practices—such as by providing projects with sufficient cost and schedule reserves to address risks and unforeseen technical challenges and ensuring that program offices regularly and consistently update their JCLs across the portfolio— could help to better position NASA for improved outcomes. We look forward to continuing to work with NASA and this subcommittee in addressing these issues.

Chairman Babin, Ranking Member Bera, and Members of the Subcommittee, this completes my prepared statement. I would be pleased to respond to any questions that you may have at this time.

ACKNOWLEDGMENTS

GAO staff who made key contributions to this statement include Molly Traci, Assistant Director; Laura Greifner; Erin Kennedy; Miranda Riemer; Roxanna T. Sun; and Alyssa Weir.

APPENDIX I. DESCRIPTIONS OF NATIONAL AERONAUTICS AND SPACE ADMINISTRATION MAJOR PROJECTS REVIEWED IN GAO'S 2018 ASSESSMENT

Table 4. Descriptions of the 26 National Aeronautics and Space Administration (NASA) Major Projects Reviewed in GAO's 2018 Assessment

Project name	Project description
Commercial Crew Program	The Commercial Crew Program facilitates and oversees the development of safe, reliable, and cost-effective crew transportation systems by commercial companies to carry NASA astronauts to and from the International Space Station. The program is a multi-phase effort that started in 2010. During the current phase, the program is working with two contractors—Boeing and SpaceX—that will design, develop, test, and operate the crew transportation systems. Once NASA determines the systems meet its standards for human spaceflight—a process called certification—the companies will fly up to six crewed missions to the space station.
Double Asteroid Redirection Test (DART)	The DART project plans to travel to the near-Earth asteroid Didymos, a binary system, and impact the smaller of the two bodies. NASA will assess the deflection result of the impact for potential future use on other potentially hazardous near-Earth objects. The project responds to near-

Project name	Project description
	Earth object guidance by the Office of Science and Technology Policy to better understand our impact mitigation posture, and to recommendations by the National Research Council Committee to conduct a test of a kinetic impactor. The DART mission is part of the Asteroid Impact and Deflection Assessment, which is an international collaboration with the European Space Agency.
Europa Clipper	The Europa Clipper mission aims to investigate whether the Jupiter moon could harbor conditions suitable for life. The project plans to launch a spacecraft in the 2020s, place it in orbit around Jupiter, and conduct a series of investigatory flybys of Europa. The mission's planned objectives include characterizing Europa's ice shell and any subsurface water, analyzing the composition and chemistry of its surface and ionosphere, understanding the formation of its surface features, and surveying sites for a potential landed mission.
Exploration Ground Systems (EGS)	The EGS program is modernizing and upgrading infrastructure at the Kennedy Space Center and developing software needed to integrate, process, and launch the Space Launch System (SLS) and Orion Multi-Purpose Crew Vehicle (Orion). The EGS program consists of several major construction and facilities projects including the Mobile Launcher, Crawler Transporter, Vehicle Assembly Building, and launch pad, all of which need to be complete before the first uncrewed exploration mission using the SLS and Orion vehicles.
Gravity Recovery and Climate Experiment Follow-On (GRACE-FO)	The GRACE-FO mission will continue and expand upon the 2002 GRACE mission, which ended science operations in October 2017. The system, which consists of two spacecraft working together to obtain scientific measurements, will provide high-resolution models of Earth's gravity field and insight into water movement on and beneath the Earth's surface for up to 5 years. These models will provide rates of ground water depletion and polar ice melt and enable improved planning for droughts and floods. GRACE-FO is a collaborative effort with the German Research Centre for Geosciences.
Ice, Cloud, and Land Elevation Satellite-2 (ICESat-2)	The ICESat-2 mission is a follow-on mission to ICESat that will measure changes in polar ice-sheet mass and elevation. The measurements will provide researchers a better understanding of the mechanisms that drive polar ice changes and their effect on global sea level. ICESat-2's upgraded laser instrument will allow the satellite to make more frequent measurements and provide better elevation estimates over certain types of terrain than ICESat.
Interior Exploration using Seismic Investigations, Geodesy, and Heat Transport (InSight)	InSight is a Mars lander with two primary objectives. It is intended to further understanding of the formation and evolution of terrestrial planets by determining Mars's size, its composition, and the physical state of the core; the thickness of the crust; and the composition and structure of the mantle, as well as the thermal state of the interior. It will also determine the present level of tectonic activity and the meteorite impact rate on Mars. InSight is based on the Phoenix lander design. Phoenix successfully landed on Mars in 2008.

Table 4. (Continued)

Project name	Project description
Ionospheric Connection Explorer (ICON)	The ICON observatory will orbit Earth to explore its ionosphere—the boundary region between Earth and space where ionized plasma and neutral gas collide and react. Its four instruments will make direct measurements and use remote sensing to further researchers' understanding of Earth's upper atmosphere, the Earth-Sun connection, and the ways in which Earth weather drives space weather.
James Webb Space Telescope (JWST)	JWST is a large, infrared-optimized space telescope designed to help understand the origin and destiny of the universe, the creation and evolution of the first stars and galaxies, and the formation of stars and planetary systems. It will also help further the search for Earth-like planets. JWST will have a large primary mirror composed of 18 smaller mirrors and a sunshield the size of a tennis court. Both the mirror and sunshield are folded for launch and open once JWST is in space. JWST will reside in an orbit about 1 million miles from the Earth.
Landsat 9	Landsat 9 is the next satellite in the Landsat series Program, which provides a continuous space-based record of land surface observations to study, predict, and understand the consequences of land surface dynamics, such as deforestation. The program is a collaborative, joint mission between NASA and the U.S. Geological Survey. The Landsat data archive constitutes the longest continuous moderate-resolution record of the global land surface as viewed from space and is used by many fields, such as agriculture, mapping, forestry, and geology.
Laser Communications Relay Demonstration (LCRD)	LCRD is a technology demonstration mission with the goal of advancing optical communication technology for use in deep space and near-Earth systems. LCRD will demonstrate bidirectional laser communications between a satellite and ground stations, develop operational procedures, and transfer the technology to industry for future use on commercial and government satellites. NASA anticipates using the technology as a next generation Earth relay as well as to support near-Earth and deep space science, such as the International Space Station and human spaceflight missions. The project is a mission partner and will be a payload on a U.S. Air Force Space Test Program satellite.
Low Boom Flight Demonstrator (LBFD)	LBFD is a flight demonstration project planned to demonstrate that noise from supersonic flight—sonic boom—can be reduced to acceptable levels, allowing for eventual commercial use of overland supersonic flight paths. Plans include multiple flights beyond fiscal year 2022 to gather community responses to the flights and to create a database to support development of international noise rules for supersonic flight.
Lucy	Lucy will be the first mission to investigate the Trojans, which are a population of never explored asteroids orbiting in tandem with Jupiter. The project aims to understand the formation and evolution of planetary systems by conducting flybys of these remnants of giant planet formation. The Lucy spacecraft will first encounter a main belt asteroid—located between the orbits of Mars and Jupiter—and then will travel to the outer solar system where the spacecraft will encounter six

Project name	Project description
	Trojans over an 11-year mission. The mission's planned measurements include asteroid surface color and composition, interior composition, and surface geology.
Mars 2020	Mars 2020 is part of the Mars Exploration Program, which seeks to further understand whether Mars was, is, or can be a habitable planet. Its rover and science instruments will explore Mars and conduct geological assessments, search for signs of ancient life, determine potential environmental habitability, and prepare soil and rock samples for potential future return to Earth. The rover will include a technology demonstration instrument designed to convert carbon dioxide into oxygen. Mars 2020 is based heavily on the Mars Science Laboratory, or Curiosity, which landed on Mars in 2012 and remains in operation.
NASA ISRO Synthetic Aperture Radar (NISAR)	NISAR is a joint project between NASA and Indian Space Research Organisation (ISRO) that will study the solid Earth, ice masses, and ecosystems. It aims to address questions related to global environmental change, Earth's carbon cycle, and natural hazards, such as earthquakes and volcanoes. The project will include the first dual frequency synthetic aperture radar instrument, which will use advanced radar imaging to construct large-scale data sets of the Earth's movements. NISAR represents the first major aerospace science partnership between NASA and ISRO.
Orion Multi-Purpose Crew Vehicle (Orion)	Orion is being developed to transport and support astronauts beyond low-Earth orbit, including traveling to Mars or an asteroid. The Orion program is continuing to advance development of the human safety features, designs, and systems started under the Constellation program, which was canceled in 2010. Orion is planned to launch atop NASA's Space Launch System. The current design of Orion consists of a crew module, service module, and launch abort system.
Parker Solar Probe (PSP)	PSP will be the first NASA mission to visit a star. Using the gravity of Venus, the spacecraft will orbit the Sun 24 times and gather information to increase knowledge about the solarwind, including its origin, acceleration, and how it is heated. PSP instruments will observe the generation and flow of solar winds from very close range and sample and take measurements of the Sun's outer atmosphere, where solar particles are energized. To achieve its mission, parts of the spacecraft must be able to withstand temperatures exceeding 2,500 degrees Fahrenheit and endure blasts of extreme radiation. The project was formerly named Solar Probe Plus, or SPP, and was renamed in May 2017.
Plankton, Aerosol, Cloud, ocean Ecosystem (PACE)	PACE is a polar-orbiting mission that will use advanced global remote sensing instruments to improve scientists' understanding of ocean biology, biogeochemistry, ecology, aerosols, and cloud properties. PACE will extend climate-related observations begun under earlier NASA missions, which will enable researchers to study long-term trends on Earth's oceans and atmosphere, and ocean-atmosphere interactions. PACE will also enable assessments of air and coastal water quality, such as the locations of harmful algae blooms.

Table 4. (Continued)

Project name	Project description
Psyche	Psyche will be the first mission to visit a metal asteroid and aims to understand a previously unexplored component of the early building blocks of planets: iron cores. The project plans to orbit the Psyche asteroid to determine if it is a planetary core, characterize its topography, assess the elemental composition, and determine the relative ages of its surface regions.
Radiation Budget Instrument (RBI)	RBI is a scanning radiometer that NASA planned to launch on the National Oceanic and Atmospheric Administration's (NOAA) Joint Polar Satellite System 2. RBI's planned mission was to support global climate monitoring by continuing measurements of the Earth's reflected sunlight and emitted thermal radiation made by NASA and NOAA satellites over the past 30 years. This data was intended to represent one of two key sets of measurements needed to determine whether the Earth is warming or cooling.
Restore-L	The Restore-L project will demonstrate the capability to refuel on-orbit satellites for eventual use by commercial entities. Specifically, Restore-L plans to autonomously rendezvous with, inspect, capture, refuel, adjust the orbit of, safely release, and depart from the U.S. Geological Survey's Landsat 7 satellite. Landsat 7 can extend operations if successfully refueled, but it is planned for retirement if the technology demonstration is unsuccessful.
Space Launch System (SLS)	SLS is intended to be NASA's first human-rated heavy-lift launch vehicle since the Saturn V was developed for the Apollo program. SLS is planned to launch NASA's Orion spacecraft and other systems on missions between the Earth and Moon and to enable deep space missions, including Mars. NASA is designing SLS to provide an initial lift capacity of 70 metric tons to low-Earth orbit, and be evolvable to 130 metric tons, enabling deep space missions. The 70-metric-ton capability will include a core stage, powered by four RS-25 engines, and two five-segment boosters. The 130-metric-ton capability will use a new upper stage and evolved boosters.
Space Network Ground Segment Sustainment (SGSS)	The SGSS project plans to develop and deliver a new ground system for one Space Network site. The Space Network provides essential communications and tracking services to NASA and non-NASA missions. Existing systems, based on 1980s technology, are increasingly obsolete and unsustainable. The new ground system will include updated systems, software, and equipment that will allow the Space Network to continue to provide critical communications services for the next several decades. The Space Network is managed by the Space Communication and Navigation program.
Surface Water and Ocean Topography (SWOT)	The SWOT mission will use its wide-swath radar altimetry technology to take repeated high resolution measurements of the world's oceans and freshwater bodies to develop a global survey. This survey will make it possible to estimate water discharge into rivers more accurately, and help improve flood prediction. It will also provide global measurements of

Project name	Project description
	ocean surface topography and variations in ocean currents, which will help improve weather and climate predictions. SWOT is a joint project between NASA and the French Space Agency—the Centre National d'Etudes Spatiales.
Transiting Exoplanet Survey Satellite (TESS)	TESS will use four identical, wide field-of-view cameras to conduct the first extensive survey of the sky from space for transiting exoplanets—or planets in other solar systems. The mission's goal is to discover these exoplanets during transit, the time when the planet's orbit carries it in front of its star as viewed from Earth. The project plans to discover rocky and potentially habitable Earth-sized and super-Earth planets orbiting nearby bright stars for further evaluation through ground- and space-based observations by other missions, such as JWST.
Wide-Field Infrared Survey Telescope (WFIRST)	WFIRST is an observatory designed to perform wide-field imaging and survey of the nearinfrared sky to answer questions about the structure and evolution of the universe, and expand our knowledge of planets beyond our solar system. The project will use a telescope that was originally built and qualified by another federal agency. The project plans to launch WFIRST in the mid-2020s to an orbit about 1 million miles from the Earth. The project is also planning a guest observer program, in which the project may provide observation time to academic and other institutions.

Source: GAO. | GAO-18-576T.

In: NASA: Major Projects and Select Issues
Editor: Oliver Cummings
ISBN: 978-1-53614-817-6

Chapter 3

NASA: FY2017 BUDGET AND APPROPRIATIONS*

Daniel Morgan

SUMMARY

The National Aeronautics and Space Administration (NASA) was created in 1958 by the National Aeronautics and Space Act (P.L. 85-568) to conduct civilian space and aeronautics activities. It has four mission directorates. The Science Mission Directorate manages robotic science missions, such as the Hubble Space Telescope, the Mars rover Curiosity, and satellites for Earth science research. The Aeronautics Research Mission Directorate conducts research and development on aircraft and aviation systems. The Space Technology Mission Directorate develops technologies for use in future space missions, such as advanced propulsion and laser communications. The Human Exploration and Operations Mission Directorate is responsible for human spaceflight activities, including the International Space Station and development efforts for future crewed spacecraft. In addition, NASA's Office of Education manages formal and informal education programs for school children, college and university students, and the general public.

While Congress is generally supportive of most NASA programs, government-wide fiscal constraints make funding decisions challenging. The Administration has requested \$19.025 billion for NASA in FY2017. This amount is 1.3% less than the FY2016 appropriation of \$19.285 billion. Unusually, the FY2017 request includes \$763 million in mandatory funds. The House bill (H.R. 5393) would provide \$19.508 billion. The Senate bill (S. 2837) would provide \$19.306 billion. Neither bill includes mandatory funding.

The FY2017 request for the Science Mission Directorate is \$5.601 billion, an increase of 0.2% from FY2016. Within this total, funding for Earth Science, Astrophysics, and Heliophysics would increase, while funding for Planetary Science and the James Webb Space Telescope would decrease. The House bill would provide \$5.597 billion for Science, while the Senate bill would provide \$5.395 billion. Within these totals, the bills differ considerably in their allocation of funding between Earth Science and Planetary Science.

* This is an edited, reformatted and augmented version of Congressional Research Service, Publication No. R44397, dated December 13, 2016.

The FY2017 request for the Aeronautics Research Mission Directorate is $790 million, an increase of 23.5% from FY2016. The request includes New Aviation Horizons (NAH), a new initiative of experimental aircraft and systems demonstrations. The House and Senate bills would provide $712 million and $601 million, respectively, for Aeronautics.

The FY2017 request for the Space Technology Mission Directorate is $827 million, an increase of 20.4% from FY2016. The House and Senate bills would provide $739 million and $687 million.

For the Human Exploration and Operations Mission Directorate, the FY2017 request for Exploration is $3.337 billion, a decrease of 17.2% from FY2016, while the request for Space Operations is $5.076 billion, an increase of 0.9%. The Exploration request includes $1.263 billion, a decrease of 35.2%, for Space Launch System launch vehicle development. Funding for the Commercial Crew program (formerly requested in Exploration) is combined with funding for operational cargo and crew transport to the International Space Station in a new Space Transportation item within Space Operations. The House bill would provide $4.183 billion for Exploration, including $2.000 billion for SLS development, and $4.890 billion for Space Operations. The Senate bill would provide $4.330 billion for Exploration, including $2.150 billion for the SLS, and $4.951 billion for Space Operations.

The FY2017 request for the Office of Education is $100 million, a decrease of 13.0% from FY2016. The House and Senate bills would provide $115 million and $108 million, respectively. The request would reduce funding for the National Space Grant College and Fellowship Program, the Experimental Program to Stimulate Competitive Research, and the Minority University Research Education Program. Both bills would fund these programs at their FY2016 levels.

The National Aeronautics and Space Administration (NASA) was created in 1958 by the National Aeronautics and Space Act (P.L. 85-568) to conduct civilian space and aeronautics activities. It has four mission directorates. The Science Mission Directorate manages robotic science missions, such as the Hubble Space Telescope, the Mars rover Curiosity, and satellites for Earth science research. The Aeronautics Research Mission Directorate conducts research and development on aircraft and aviation systems. The Space Technology Mission Directorate develops new technologies for use in future space missions, such as advanced propulsion and laser communications. The Human Exploration and Operations Mission Directorate is responsible for human spaceflight activities, including the International Space Station and development efforts for future crewed spacecraft. In addition, NASA's Office of Education manages formal and informal education programs for school children, college and university students, and the general public.

While Congress is generally supportive of most NASA programs, government-wide fiscal constraints make funding decisions challenging. This report presents an overview and selected highlights of NASA's proposed FY2017 budget and the FY2017 appropriations legislation reported by the House Committee on Appropriations (H.R. 5393, H.Rept. 114-605) and the Senate Committee on Appropriations (S. 2837, S.Rept. 114-239). Pending final action on FY2017 appropriations for NASA, the Further Continuing and Security Assistance Appropriations Act, 2017 (P.L. 114-254) provides continuing appropriations at 99.8% of the FY2016 level through April 28, 2017. This report will be updated as Congress takes additional action on FY2017 appropriations legislation.

OVERVIEW

The Administration has requested $19.025 billion for NASA in FY2017. This amount is 1.3% less than the FY2016 appropriation of $19.285 billion.[1] The House bill would provide $19.508 billion. The Senate bill would provide $19.306 billion. See *Table 1.*

Unusually, the FY2017 request includes $763 million in mandatory funds. The Administration explains this as follows:

> The President believes that arbitrary funding caps are harmful to the economy and the Nation. Although the recent Bipartisan Budget Act provided important relief from sequester cuts, the constrained top line for discretionary funding, especially in FY 2017, has made it difficult to appropriately fund important national priorities, including research and development. In the FY 2017 Budget, the President proposes fully-paid-for one-year mandatory funding to accelerate progress in Science, Aeronautics, Space Technology, and Exploration, as well as additional multi-year mandatory funding for Aeronautics to support research and development for low carbon emission aircraft, including associated transportation systems, as part of a multi-agency effort to enable a 21st century clean transportation system.[2]

Typically, NASA receives no more than a few million dollars per year of mandatory funding. In FY2015, for example, it received $2 million for Space Operations from the Spectrum Reallocation Fund and $1 million from the Science, Space, and Technology Education TrustFund.[3] The discretionary portion of the FY2017 request, $18.262 billion, would be 5.3% less than the FY2016 appropriation. The House and Senate bills include no mandatory funding.

There is no authorized level for NASA appropriations in FY2017. The most recent authorization act (the NASA Authorization Act of 2010, P.L. 111-267) authorized appropriations through FY2013. The NASA Authorization Act for 2016 and 2017 (H.R. 2039, ordered reported by the House Committee on Science, Space, and Technology in April 2015, but not yet reported) and the NASA Transition Authorization Act of 2016 (S. 3346, passed by the Senate on December 10, 2016) include proposed authorization amounts for FY2017.[4]

---

[1] FY2016 amounts in this chapter are based on the Consolidated Appropriations Act, 2016 (P.L. 114-113) and the accompanying explanatory statement, *Congressional Record*, December 17, 2015, at pp. H9741-H9743. Some amounts (but not the total) may change under NASA operating plans that reflect transfers and reprogramming.

[2] FY2017 NASA congressional budget justification, p. SUM-10. The justification does not identify a source for most of the proposed mandatory funding.

[3] FY2017 Budget of the United States Government, Appendix, pp. 1190 and 1193.

[4] For details, see CRS Report R43419, *NASA Appropriations and Authorizations: A Fact Sheet*, by Daniel Morgan.

Table 1. NASA Appropriations, FY2016-FY2017 (budget authority in $ millions)

	FY2016	FY2017 Request			FY2017 Legislation	
	Enacted	Total	Mand.	Discr.	H. Cte.	S. Cte. Enacted
Science	$5,589	$5,601	$298	$5,303	$5,597	5,395
Earth Science	1,921	2,032	60	1,972	1,690	1,984
Planetary Science	1,631	1,519	128	1,391	1,846	1,356
Astrophysics	731	782	85	697	793	807
James Webb Space Telescope	620	569	—	569	569	569
Heliophysics	650	699	25	674	699	679
Education	37	—[a]	—	—[a]	—[b]	—[c]
Aeronautics	640	790	156	635	712	601
Space Technology	687	827	136	691	739	687
Exploration	4,030	3,337	173	3,164	4,183	4,330
Exploration Systems Development	3,680	2,860	173	2,687	3,779	3,934
- Orion MPCV	*1,270*	*1,120*	*66*	*1,053*	*1,350*	*1,300*
- Space Launch System	*2,000*	*1,310*	*80*	*1,230*	*2,000*	*2,150*
- Exploration Ground Systems	*410*	*429*	*26*	*403*	*429*	*484*
Exploration R&D	350	477	—	477	404	396
Space Operations	5,029	5,076	—	5,076	4,890	4,951
Space Shuttle	n/s	—	—	—	n/s	8
International Space Station	n/s	1,431	—	1,431	n/s	n/s
Space Transportation	n/s[d]	2,758[e]	—	2,758[e]	n/s	n/se
Space and Flight Support	n/s	887	—	887	n/s	n/s
Education	115	100	—	100	115	108
Space Grant	40	24	—	24	40	40
EPSCoR	18	9	—	9	18	18
MUREP	32	30	—	30	32	32
Other	25	37	—	37	25	18
Safety, Security, & Mission Services	2,769	2,837	—	2,837	2,835	2,797
Construction & EC&R	389	420	—	420	398	400
Inspector General	37	38	—	38	38	38
Total	19,285	19,025	763	18,262	19,508	19,306

Sources: FY2016 enacted from P.L. 114-113 and pp. H9741-H9743 of the explanatory statement, Congressional Record, December 17, 2015. FY2017 request from NASA FY2017 congressional budget justification. FY2017 House Committee from H.R. 5393 as reported and H.Rept. 114-605. FY2017 Senate Committee from S. 2837 as reported and S.Rept. 114-239.

Notes: Some totals may not add because of rounding. For readability, a dash indicates zero. n/s = not specified. MPCV = Multipurpose Crew Vehicle. R&D = Research and Development. EPSCoR = Experimental Program to Stimulate Competitive Research. MUREP = Minority University Research Education Program. EC&R = Environmental Compliance and Remediation.

[a] Astrophysics includes $25 million for education.

[b] Other Science themes include $37 million for education and public outreach.

[c] Astrophysics includes $42 million for education.

[d] Includes up to $1,244 million for Commercial Crew.

[e] Includes $1,185 million for Commercial Crew.

SCIENCE

The FY2017 request for the Science Mission Directorate is $5.601 billion, an increase of 0.2% from FY2016. Within this total, funding for Earth Science, Astrophysics, and Heliophysics would increase, while funding for Planetary Science and the James Webb Space Telescope would decrease. As shown in *Table 1*, the House bill would provide $4 million less than the request, while the Senate bill would provide $206 million less. Relative to the request, the House bill would shift more than $300 million from Earth Science to Planetary Science. In contrast, most of the Senate bill's decrease would be in Planetary Science.

Within Earth Science, the request includes $131 million (up from $100 million in FY2016) for the Landsat-9 land imaging satellite. Launch is anticipated "as early as" 2021. NASA previously proposed the Thermal Infrared Free Flyer, a lower-cost satellite intended to reduce the risk of a gap in data availability prior to the launch of Landsat-9. Congress rejected funding for the Thermal Infrared Free Flyer in the FY2016 appropriations cycle, and the mission is not included in the FY2017 request. The House report directs NASA to prioritize funds for Landsat-9 and to evaluate commercially available data in the event of a data gap in the Landsat program. The Senate report recommends the requested amount for Landsat-9 and directs NASA to provide a plan detailing the technical and schedule progress needed for a 2020 launch date.

Within Planetary Science, the request includes $50 million (down from $175 million in FY2016) for a mission to Jupiter's moon Europa. Although a mission to Europa was a high priority of the 2011 National Research Council (NRC) decadal survey of planetary science,[5] the NRC expressed reservations about its anticipated cost. For several years, Congress has appropriated more for formulation of a Europa mission than NASA has requested. As directed by the Consolidated Appropriations Act, 2016, NASA's FY2017 congressional budget justification includes a five-year estimate of the funding required assuming a 2022 launch (see *Table 2*). The justification states that "the notional outyear profile in the Budget may support a launch as early as the late 2020s, assuming the mission concept and scope remain stable.... Acceleration of the launch to 2022 is not recommended, given potential impacts to the rest of the Science portfolio." The House report recommends at least $260 million for Europa orbiter and lander missions, with the orbiter launch no later than 2022 and the lander launch no later than 2024. The Senate report calls for "an expeditious launch and reduced travel time" in order to maximize the scientific return of a Europa mission, but it does not specify a funding level or a launch date. It directs NASA to provide a report on options for the mission "to assist the Committee in evaluating potential mission configurations."

---

[5] National Research Council, *Vision and Voyages for Planetary Science in the Decade 2013-2022* (National Academies Press, 2011). Available online at http://www.nap.edu/catalog.php?record_id=13117.

AERONAUTICS

The FY2017 request for the Aeronautics Research Mission Directorate is $790 million, an increase of 23.5% from FY2016. The request includes New Aviation Horizons (NAH), a new initiative of experimental aircraft and systems demonstrations. NAH projects on subsonic aircraft would receive $100 million in mandatory funding from the President's proposed 21st Century Clean Transportation Plan.[6] An additional $56 million in mandatory funding would fund a low-boom supersonic flight demonstrator.

As shown in *Table 1*, the House bill would provide $78 million less than the request for Aeronautics, while the Senate bill would provide $189 million less. The House committee's recommendation includes $61 million (of discretionary funds) for a low-boom flight demonstrator. The House and Senate committee reports both direct NASA to work with the Federal Aviation Administration on research related to the integration of unmanned aerial systems in the National Airspace System.[7]

Table 2. Estimated Five-Year Cost of a Europa Mission Assuming Launch in 2022 ($ in millions)

FY2017	FY2018	FY2019	FY2020	FY2021
$194	$272	$456	$678	$482

Source: NASA FY2017 congressional budget justification, p. PS-58.

SPACE TECHNOLOGY

The FY2017 request for the Space Technology Mission Directorate is $827 million, an increase of 20.4% from FY2016. Space Technology was first established as a separate account in FY2011. Each year since then, the Administration has proposed to increase Space Technology funding. Congress has provided increases each year except FY2014, but always less than the Administration's request. Proposed mandatory funding of $136 million would account for almost all of the requested increase in FY2017. The bulk of the mandatory funding would support technology demonstration missions, including the Restore-L satellite servicing mission for which Congress appropriated $133 million in FY2016.

---

[6] See "Fact Sheet: President Obama's 21st Century Clean Transportation System," White House press release, February 4, 2016, https://www.whitehouse.gov/the-press-office/2016/02/04/fact-sheet-president-obamas-21st-century-cleantransportation-system.

[7] For more information on this issue, see CRS Report R44352, *Unmanned Aircraft Operations in Domestic Airspace: U.S. Policy Perspectives and the Regulatory Landscape*, by Bart Elias.

As shown in *Table 1*, the House bill would provide $88 million less than the request for Space Technology, but more than the FY2016 appropriation, while the Senate bill would provide the FY2016 amount. The Senate committee report recommends $130 million for Restore-L. The House and Senate committee reports both identify nuclear propulsion research and a small launch technology demonstration platform as funding priorities.

HUMAN EXPLORATION AND OPERATIONS

The Human Exploration and Operations Mission Directorate (HEOMD) is funded by two appropriations accounts: Exploration and Space Operations. The FY2017 request for Exploration is $3.337 billion, a decrease of 17.2% from FY2016. The request for Space Operations is $5.076 billion, an increase of 0.9%. As shown in *Table 1*, the House and Senate bills would both provide significantly more than the request for Exploration and somewhat less than the request for Space Operations.

Exploration

The Exploration account primarily funds development of the Orion Multipurpose Crew Vehicle and the Space Launch System (SLS) heavy-lift rocket, the capsule and launch vehicle mandated by the NASA Authorization Act of 2010 for future human exploration beyond Earth orbit. The account previously also funded development of a commercial crew transportation capability for U.S. astronaut access to the International Space Station (ISS), but Congress transferred this activity to Space Operations in FY2016. Within Exploration, the FY2017 request for Orion, the SLS, and related ground systems (known collectively as Exploration Systems Development) is $2.860 billion, a decrease of 22.3% from FY2016. The bulk of the reduction would be for SLS launch vehicle development, which would receive $1.263 billion, down 35.2% from $1.950 billion in FY2016. According to NASA, the SLS program remains on track for a first test flight carrying Orion but no crew (known as EM-1) in November 2018. The launch readiness date for the first flight of Orion and the SLS with a crew on board (known as EM-2) continues to be FY2023.

The increases for Exploration in the House and Senate bills, relative to the request, would fund the SLS at the FY2016 level (in the House bill) or higher (in the Senate bill). The House and Senate reports both recommend funding for development of the SLS Exploration Upper Stage (EUS)—$250 million and $300 million, respectively—which is not included in the Administration request. The House report specifies that none of the funds in the House bill are for the Asteroid Redirect Mission, which was first proposed in the FY2014 budget and which has faced ongoing opposition in Congress.

Space Operations

The Space Operations account primarily funds operational human spaceflight: the ISS, transportation of crew and cargo to and from the ISS, and formerly the space shuttle program.[8] It also funds the Space and Flight Support program, which consists of launch site operations, space-to-ground communications, and other support activities. Because Congress gave limited direction about how the FY2016 appropriation for Space Operations should be allocated, it is difficult to determine how the allocation of the FY2017 request compares to FY2016. The FY2017 budget is the first to highlight Space Transportation (to and from the ISS) as a top-level item within Space Operations. In previous budgets, the cost of U.S. commercial cargo flights to the ISS and payments to Russia for Soyuz flights carrying ISS crews were subcategories within the ISS budget. Meanwhile, the program to develop a U.S. capability for commercial transportation of ISS crews was funded in the Exploration account. These functions are now combined in the request for Space Transportation.

The House and Senate bills would provide $186 million less and $125 million less, respectively, than the request for Space Operations. For the most part, the committee reports do not specify how the recommended funding should be allocated. However, the Senate report does note that its total includes $1.185 billion for Commercial Crew, the same as the request.

Orion and SLS vs. Commercial Crew

The balance of funding between Exploration Systems Development and the Commercial Crew program has been contentious. Before the FY2017 request, recent Administration budgets proposed to decrease funding for Exploration Systems Development while increasing funding for Commercial Crew. NASA argued that the amounts it requested for the Commercial Crew program were necessary to maintain the scheduled availability of commercial crew transportation to the ISS starting in 2017. It noted that without a U.S. commercial capability, it would need to pay Russia for additional Soyuz flights (although it has also stated that it will likely purchase some additional Soyuz flights in any case). Meanwhile, NASA officials stated that the schedule for the EM-1 test flight would be difficult to accelerate, even with additional funding, because it depends on technical requirements such as engineering design and manufacturing schedules and the need for adequate testing.[9] Congressional supporters countered that the Orion and SLS

---

[8] Although the last flight of a space shuttle was in 2011, program closeout activities, such as subcontractor audits, are still ongoing.

[9] See, for example, Charles F. Bolden, Jr., Administrator, National Aeronautics and Space Administration, testimony before the House Committee on Science, Space, and Technology, March 7, 2012.

programs were not receiving the funds they needed, and many saw this pattern as demonstrating a difference in human spaceflight priorities between Congress and the Administration. Congress has generally appropriated less than the Administration's request for Commercial Crew and more for Exploration Systems Development. In the FY2017 request, the Administration is again requesting a reduction for Exploration Systems Development, especially the SLS, but it appears to be requesting approximately the FY2016 amount for Commercial Crew (although the FY2016 allocation of Space Operations funds is not yet clear). The House and Senate bills would again provide more than the request for Exploration Systems Development, but the Senate bill would provide the requested amount for Commercial Crew. The House report does not specify a recommended allocation for Commercial Crew.

EDUCATION

The FY2017 request for the Office of Education is $100 million, a decrease of 13.0% from FY2016. Programs of particular congressional interest include the National Space Grant College and Fellowship Program ($24 million), the Experimental Program to Stimulate Competitive Research (EPSCoR, $9 million), and the Minority University Research Education Program (MUREP, $30 million). Note that not all NASA education activities are managed or funded by the Office of Education. For example, the request for Astrophysics in the Science Mission Directorate includes $25 million for science, technology, engineering, and mathematics (STEM) education. However, recent Administration initiatives to consolidate and reorganize STEM education activities across the government appear to have reduced NASA's efforts outside the Office of Education. Most notably, the previous Science Mission Directorate policy, under which 1% of all Science mission funding was allocated to education and public outreach, is no longer in effect.

As shown in *Table 1*, the House and Senate bills would both provide more than the request for the Office of Education: $115 million and $108 million, respectively. Within these totals, both would provide the FY2016 amounts for Space Grant ($40 million), EPSCoR ($18 million), and MUREP ($32 million). For education activities in the Science Mission Directorate, the House report recommends $37 million, while the Senate report recommends $42 million.

OTHER ACCOUNTS

The three remaining NASA appropriations accounts fund cross-cutting, supporting, and oversight activities. The FY2017 request for Safety, Security, and Mission Services, which funds management and operations at the NASA centers and NASA headquarters, is \$2.837 billion, an increase of 2.5% from FY2016. In part, this increase reflects the transfer of funding from other accounts to support the consolidation of NASA information technology services. The request for Construction and Environmental Compliance and Remediation is \$420 million, an increase of 7.9%. The request for the Office of Inspector General is \$38 million, an increase of 1.9%. The House and Senate recommendations for these accounts are shown in *Table 1*.

In: NASA: Major Projects and Select Issues
Editor: Oliver Cummings
ISBN: 978-1-53614-817-6

Chapter 4

NASA INFORMATION TECHNOLOGY: URGENT ACTION NEEDED TO ADDRESS SIGNIFICANT MANAGEMENT AND CYBERSECURITY WEAKNESSES*

United States Government Accountability Office

ABBREVIATIONS

CIO	Chief Information Officer
IT	information technology
NASA	National Aeronautics and Space Administration
NIST	National Institute for Standards and Technology
OMB	Office of Management and Budget
OPM	Office of Personnel Management
SAISO	Senior Agency Information Security Officer

WHY GAO DID THIS STUDY

NASA depends heavily upon IT to conduct its work. The agency spends at least $1.5 billion annually on IT investments that support its missions, including ground control systems for the International Space Station and space exploration programs.

* This is an edited, reformatted and augmented version of United States Government Accountability Office Report to Congressional Committees, Accessible Version. Publication No. GAO-18-337, May 2018.

The National Aeronautics and Space Administration Transition Authorization Act of 2017 included a provision for GAO to review the effectiveness of NASA's approach to overseeing and managing IT, including its ability to ensure that resources are aligned with agency missions and are cost effective and secure. Accordingly, GAO's specific objective for this review was to determine the extent to which NASA has established and implemented leading IT management practices in strategic planning, workforce planning, governance, and cybersecurity. To address this objective, GAO compared NASA IT policies, strategic plans, workforce gap assessments, and governance board documentation to federal law and leading practices. GAO also assessed NASA IT security plans, policies, and procedures against leading cybersecurity risk management practices.

WHAT GAO RECOMMENDS

GAO is making 10 recommendations to NASA to address the deficiencies identified in NASA IT strategic planning, workforce planning, governance, and cybersecurity. NASA concurred with seven recommendations, partially concurred with two, and did not concur with one. GAO maintains that all of the recommendations discussed in this report remain valid.

WHAT GAO FOUND

The National Aeronautics and Space Administration (NASA) has not yet effectively implemented leading practices for information technology (IT) management. Specifically, GAO identified weaknesses in NASA's IT management practices for strategic planning, workforce planning, governance, and cybersecurity.

- NASA has not documented its IT strategic planning processes in accordance with leading practices. While NASA's updated IT strategic plan represents improvement over its prior plan, the updated plan is not comprehensive because it does not fully describe strategies for achieving desired results or describe interdependencies within and across programs. Until NASA establishes a comprehensive IT strategic plan, it will lack critical information needed to align resources with business strategies and investment decisions.
- Of the eight key IT workforce planning activities, the agency partially implemented five and did not implement three. For example, NASA does not assess competency and staffing needs regularly or report progress to agency

leadership. Until NASA implements the key IT workforce planning activities, it will have difficulty anticipating and responding to changing staffing needs.

- NASA's IT governance does not fully address leading practices. While the agency revised its governance boards, updated their charters, and acted to improve governance, it has not fully established the governance structure, documented improvements to its investment selection process, fully implemented investment oversight practices and ensured the Chief Information Officer's visibility into all IT investments, or fully defined policies and procedures for IT portfolio management. Until NASA addresses these weaknesses, it will face increased risk of investing in duplicative investments or may miss opportunities to ensure investments perform as intended.

NASA has not fully established an effective approach to managing agency-wide cybersecurity risk. An effective approach includes establishing executive oversight of risk, a cybersecurity risk management strategy, an information security program plan, and related policies and procedures.

As NASA continues to collaborate with other agencies and nations and increasingly relies on agreements with private companies to carry out its missions, the agency's cybersecurity weaknesses make its systems more vulnerable to compromise. Until NASA leadership fully addresses these leading practices, its ability to ensure effective management of IT across the agency and manage cybersecurity risks will remain limited.

NASA Implementation of Cybersecurity Risk Management Practices

Practice	Status
Executive oversight of risk	While NASA has designated a risk executive, the agency lacks a dedicated office to provide comprehensive executive oversight of risks.
Cybersecurity risk management strategy	NASA lacks an agency-wide cybersecurity risk management strategy; one is currently in development.
Information security program plan	NASA developed a draft agency-wide information security program plan; however, the plan does not yet fully address leading practices.
Policies and procedures	Policies and procedures for protecting NASA's information systems are in place, but the agency has not kept them current or integrated.

Source: GAO analysis of National Aeronautics and Space Administration documentation. | GAO-18-337.

May 22, 2018

The Honorable John Thune
Chairman

The Honorable Bill Nelson
Ranking Member
Committee on Commerce, Science, and Transportation
United States Senate

The Honorable Lamar Smith
Chairman

The Honorable Eddie Bernice Johnson
Ranking Member
Committee on Science, Space, and Technology
House of Representatives

The National Aeronautics and Space Administration (NASA) exercises control over aeronautical and space activities sponsored by the United States and also seeks to encourage the fullest commercial use of space. NASA's current and planned activities span a broad range of complex and technical endeavors, including developing new capabilities to send future missions to Mars, improving the air transportation experience, and developing new space transportation systems. These activities may rely on partnerships with academic, private sector, and international organizations, including foreign space agencies.[1] In recent years, NASA also has increasingly relied on other countries and agreements with private companies to support its missions.[2]

NASA depends heavily upon information technology (IT) to conduct its work. Since fiscal year 2016, the agency has planned to spend about $1.5 billion annually on IT investments that support its missions, including ground control systems for the International Space Station and other space exploration programs. In addition, the agency has invested in cloud computing, data center optimization, and IT security capabilities to support its business operations.

However, we and the NASA Office of Inspector General have reported that the agency has struggled for more than two decades to overcome its decentralized operations and

[1] These include the European Space Agency as well as the space agencies of Argentina, Brazil, China, Germany, India, Israel, Japan, Russia, South Korea, and Ukraine.

[2] For example, since the Space Shuttle was retired in 2011, the United States has been relying on Russia to carry astronauts to and from the International Space Station. NASA's Commercial Crew Program is facilitating private development of a domestic system to meet that need safely, reliably, and cost-effectively before the seats it has contracted for on a Russian spacecraft run out in 2019.

culture of autonomy at its major organizational units, in an attempt to provide effective oversight and management of its IT environment. Moreover, we and others have reported on information security concerns. NASA systems are highly interconnected.[3] The agency is also affected by geopolitics and is targeted by cybercriminals, including those that may be sponsored by foreign intelligence services.[4] In addition, entities with whom NASA collaborates may also be targets of cybercriminals.

The National Aeronautics and Space Administration Transition Authorization Act of 2017 included a provision for us to review the effectiveness of NASA's approach to overseeing and managing IT, including its ability to ensure that resources are aligned with agency missions and are cost effective and secure.[5] Our objective for this review was to address the extent to which NASA has established and implemented leading IT management practices in strategic planning, workforce planning, governance, and cybersecurity.

To address this objective, we compared NASA's IT management policies, procedures, and documentation to criteria established by federal law and leading practices.

- For our work regarding IT strategic planning, we obtained and evaluated NASA's documentation on IT strategic planning, including its related planning guidance, agency-wide strategic plan, and IT-specific strategic plans. We analyzed its strategic planning guidance and assessed the contents of the previous and current IT strategic plans by comparing them to leading practices that we and the Office of Management and Budget (OMB) have previously identified.[6] These practices include documenting the IT strategic planning process and developing a strategic plan that defines the agency's vision and provides a road map to help align information resources with business strategies and investment decisions.
- For IT workforce planning, we reviewed documentation, including NASA's 2015 draft IT workforce plan, competencies, reported staffing data, and a workforce gap

---

[3] The Office of Inspector General has reported that the agency faces unique security challenges because of its connectivity with educational institutions, research facilities, and other organizations.

[4] During 2016 and 2017, NASA reported more than 3,000 computer security incidents related to malicious software on or unauthorized access to agency computers. These incidents included criminal enterprises seeking profit and intrusions that may have been sponsored by foreign intelligence services seeking to further their countries' objectives.

[5] National Aeronautics and Space Administration Transition Authorization Act of 2017, Pub. L. No. 115-10, §811(b), 131 Stat. 18, 59 (2017).

[6] Leading practices were identified related to strategic planning from the Office of Management and Budget, Circular No. A-11: *Preparation, Submission, and Execution of the Budget*, July 2017; OMB Circular No. A-130: *Managing Information as a Strategic Resource*, (Washington, D.C.: July 28, 2016); and OMB Memorandum M-13-09 Fiscal Year 2013 PortfolioStat Guidance: *Strengthening Federal IT Portfolio Management* (Washington, D.C.: Mar. 27, 2013). Further, our prior work related to IT strategic planning practices includes, for example, GAO, *Social Security Administration: Improved Planning and Performance Measures Are Needed to Help Ensure Successful Technology Modernization*, GAO-12-495 (Washington, D.C.: Apr. 26, 2012); and *Library of Congress: Strong Leadership Needed to Address Serious Information Technology Management Weaknesses*, GAO-15-315 (Washington, D.C.: Mar. 31, 2015).

assessment. We compared this documentation to eight key IT workforce planning activities that we previously identified and that were derived from federal law and guidance, including OMB memorandums, GAO reports, and the Office of Personnel Management's (OPM) Human Capital Framework.[7]

- Our review of IT governance involved analyzing NASA's governance board meeting minutes and briefings, charters, and policies and procedures, and comparing them to criteria as identified by GAO in the IT investment management framework.[8] Specifically, we focused on policies and procedures related to instituting investment boards, selecting investments, overseeing investments, and developing investment portfolios.
- Regarding cybersecurity, we obtained and analyzed available NASA documentation to determine the extent to which the agency had established an effective approach for managing cybersecurity risk.
  This documentation included information security policies and procedures, management briefings and reports, and a draft information security program plan. We then assessed whether NASA's approach addressed foundational cybersecurity risk management components identified in the National Institute for Standards and Technology's (NIST) guidance.[9] These components included the establishment of a risk executive function, a cybersecurity risk management strategy, an information security program plan, and current policies and procedures for relevant security controls.

In addition to assessing IT management at NASA headquarters, we reviewed such management practices at two of the agency's centers and one mission directorate, focusing on those with the largest fiscal year 2017 IT budgets.[10] In addition, we visited and reviewed IT management practices for the Goddard Space Flight Center because of the center's proximity to GAO. We also interviewed cognizant officials with responsibilities for IT management at NASA. The results of our work at the selected NASA centers and mission directorate are not generalizable to other NASA centers and mission directorates. Additional details on our objective, scope, and methodology can be found in appendix I.

We conducted this performance audit from May 2017 to May 2018 in accordance with generally accepted government auditing standards. Those standards require that we plan

---

[7] GAO, *IT Workforce: Key Practices Help Ensure Strong Integrated Program Teams; Selected Departments Need to Assess Skill Gaps*, GAO-17-8 (Washington, D.C.: Nov. 30, 2016).

[8] GAO, *Information Technology Investment Management: A Framework for Assessing and Improving Process Maturity*, GAO-04-394G (Washington, D.C.: March 2004) and OMB, *Memorandum M-13-09 Fiscal Year 2013 PortfolioStat Guidance: Strengthening Federal IT Portfolio Management* (Washington, D.C.: Mar. 27, 2013.

[9] NIST, Framework for Improving Critical Infrastructure Cybersecurity (Feb. 12, 2014).

[10] We conducted work at three of NASA's nine centers (two centers with the largest fiscal year 2017 IT budgets—Johnson Space Center in Houston, Texas and Marshall Space Flight Center in Huntsville, Alabama) and Goddard Space Flight Center, in Greenbelt, Maryland. We also reviewed IT management practices at one of NASA's four mission directorates (the Human Explorations and Operations Mission Directorate).

and perform the audit to obtain sufficient, appropriate evidence to provide a reasonable basis for our findings and conclusions based on our audit objectives. We believe that the evidence obtained provides a reasonable basis for our findings and conclusions based on our audit objective.

BACKGROUND

NASA's mission is to drive advances in science, technology, aeronautics, and space exploration to enhance knowledge, education, innovation, economic vitality, and stewardship of Earth. The NASA Administrator is responsible for leading the agency and is accountable for all aspects of its mission, including establishing and articulating its vision and strategic priorities and ensuring successful implementation of supporting policies, programs, and performance assessments.

Within NASA headquarters, the agency has four mission directorates that define its major core mission work: (1) *Aeronautics Research* conducts cutting-edge research to enable revolutionary advances in future aircraft, as well as in the airspace in which they will fly; (2) *Human Exploration and Operations* is responsible for NASA space operations, developing new exploration and transportation systems, and performing scientific research; (3) *Science* carries out the scientific exploration of Earth and space to expand the frontiers of Earth science, planetary science, and astrophysics, and (4) *Space Technology* develops revolutionary technologies through transparent, collaborative partnerships that expand the boundaries of aerospace. The agency also has a mission support directorate to manage its business needs and administrative functions, such as human capital management.

In addition to NASA headquarters in Washington, D.C., the agency is composed of nine field centers managed by NASA employees, and one federally funded research and development center that are responsible for executing programs and projects.[11] NASA centers are located throughout the country and manage projects or programs for multiple mission directorates. For example, the Goddard Space Flight Center supports various IT programs within the *Science* mission directorate, while the Johnson Space Center supports multiple programs in the *Human Exploration and Operations* mission directorate.

According to NASA documents, the agency planned to spend $1.6 billion of its fiscal year 2018 budget authority on IT.[12] Of this total, $888 million was to be used for business

---

[11] The nine field centers are (1) Glenn Research Center in Ohio, (2) the Goddard Space Flight Center in Maryland, (3) the Langley Research Center in Virginia, (4) the Kennedy Space Center in Florida, (5) the Marshall Space Flight Center in Alabama, (6) Stennis Space Center in Mississippi, (7) Johnson Space Center in Texas, (8) Armstrong Flight Research Center in California, and (9) Ames Research Center in California. In addition, the Jet Propulsion Laboratory is a federally funded research center managed for NASA by the California Institute of Technology.

[12] The planned $1.6 billion in IT budget authority represented about 8 percent of NASA's $19.1 billion total budget request for fiscal year 2018.

IT and $672.8 million was to be used for mission IT. Business IT includes the infrastructure and systems needed to support internal agency operations, such as commodity IT (e.g., e-mail and communications systems), infrastructure, IT management, administrative services, and support systems, whereas mission IT includes the technology needed to support space programs and research for the agency's mission programs. The technology that the agency uses to support its mission programs includes highly-specialized IT, defined by NASA as any equipment, system, and/or software that is used to acquire, store, retrieve, manipulate, and/or transmit data or information when the IT is embedded in a mission platform or provides a platform required for simulating, executing, or operating a mission.

Historically, NASA and its Inspector General have reported that funding for and oversight of highly-specialized IT has been decentralized among mission directorates and embedded within launch programs and other mission activities instead of being identified as IT to be managed as part of the agency's IT portfolio.[13] According to the Inspector General, the agency's decentralized funding for and oversight of IT has minimized agency-wide visibility into and oversight of NASA's spending on these systems.

NASA's IT Management and Governance Structure

The agency's Chief Information Officer (CIO) reports directly to the NASA Administrator and serves as the principal advisor to the NASA Administrator and senior officials on all matters pertaining to IT. The CIO is to provide leadership, planning, policy direction, and oversight for the management of NASA's information and systems. Toward this end, the CIO's responsibilities include developing and implementing approaches for executing the goals and outcomes in the NASA strategic plan; ensuring that the agency's human resources possess the requisite knowledge and skills in IT and information resources management; maximizing the value of NASA IT investments through an investment management process; and leading and implementing the agency's IT security program. The CIO also is responsible for developing and implementing agency-wide IT policies and processes.

NASA's CIO also is to direct, manage, and provide policy guidance and oversight of the agency's center CIOs. Each center has a CIO responsible for supporting center leadership and managing IT staff. Similarly, each mission directorate has a representative who coordinates with programs on IT-specific issues and, as needed, obtains support from the Office of the CIO. Both center CIOs and mission directorate IT representatives report to the NASA CIO and to the leadership of their respective centers and mission directorates.

[13] NASA, IT Portfolio Review Tiger Team: Final Report to Information Technology Council, (Washington, D.C.: Dec. 7, 2016) and National Aeronautics and Space Administration Office of Inspector General, NASA's Efforts to Improve the Agency's Information Technology Governance, IG-18-002, (Washington, D.C.: Oct. 19, 2017).

The CIO is supported by staff in the Office of the CIO. This office is organized into four divisions responsible for (1) IT security, (2) capital planning and governance, (3) technology and innovation, and (4) enterprise services and integration. Collectively, these divisions support NASA's approach to IT strategic and workforce planning, governance boards and practices, and cybersecurity.

In March 2017, the Office of the CIO submitted plans to establish a fifth division focused on new applications, and also to rename existing divisions to better represent the services they provide. For example, the Office of the CIO proposed that the Capital Planning and Governance Division be renamed the IT Business Management Division. As of March 2018, NASA had not yet approved or implemented the planned reorganization.

Figure 1 depicts the organization of the Office of the CIO, including relevant reporting relationships for center CIOs and mission directorate IT representatives, as of March 2018.

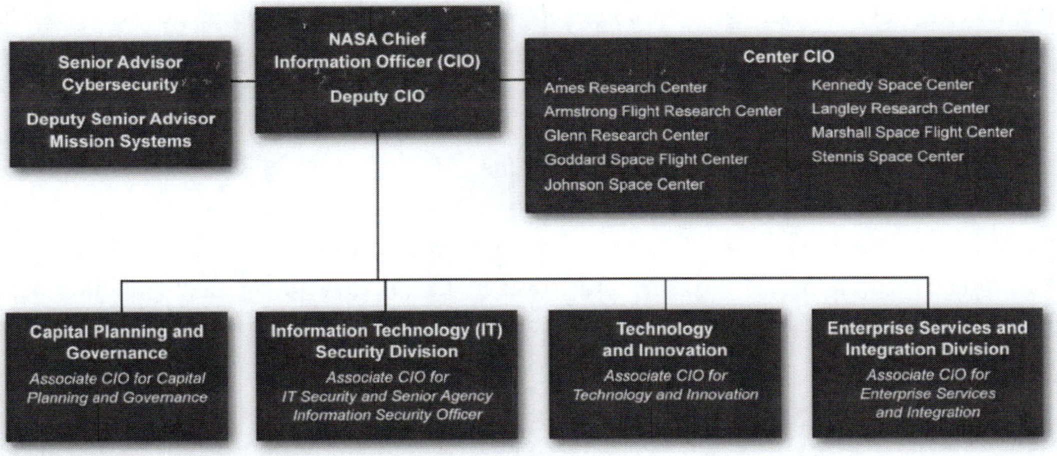

Source: GAO analysis of NASA data. | GAO-18-337.

Figure 1. Organization of the National Aeronautics and Space Administration's (NASA) Office of the Chief Information Officer, as of March 2018.

GAO and NASA's Office of Inspector General Have Reported on Longstanding Weaknesses in IT Management

We and NASA's Office of Inspector General have reported on longstanding IT management weaknesses within the agency. For example, in October 2009, we reported that NASA had made progress in implementing IT security controls and aspects of its information security program, but that it had not always implemented appropriate controls to sufficiently protect the confidentiality, integrity, and availability of information and systems.[14] We also identified control vulnerabilities and program shortfalls, which, collectively, increased the risk of unauthorized access to NASA's sensitive information, as well as inadvertent or deliberate disruption of its system operations and services. We recommended that the NASA Administrator take steps to mitigate control vulnerabilities and fully implement a comprehensive information security program. The agency concurred with our eight recommendations and stated that it was taking actions to mitigate the information security weaknesses identified.

In addition, NASA's Office of Inspector General has issued 24 reports over the last 7 years on IT governance and security weaknesses at the agency. For example, in June 2013, the office reported that the decentralized nature of NASA's operations and its longstanding culture of autonomy had hindered the agency's ability to implement effective IT governance.[15] Specifically, the report stated that the CIO had limited visibility and control over a majority of IT investments, operated in an organizational structure that marginalized the authority of the position, and could not enforce security measures across NASA's computer networks. Moreover, the IT governance structure in place at the time was overly complex, did not function effectively, and operated under a decentralized model that relegated decision making about critical IT issues to numerous individuals across NASA, leaving such decisions outside the purview of the CIO.

The Office of Inspector General made eight recommendations to the NASA Administrator for improving IT governance, including calling for all governance to be consolidated within the Office of the CIO to ensure adequate visibility, accountability, and integration into all mission-related IT assets and activities. The Administrator concurred with six and partially concurred with two of the recommendations and planned actions sufficient for the Office of Inspector General to close all eight recommendations as implemented. However, the Office of Inspector General later reported that the extent to which NASA had implemented the agreed-upon changes was in doubt based on subsequent audit findings that NASA was still struggling with limited agency CIO authority, decentralized IT operations, and ineffective IT governance.

A follow-on report issued in October 2017 described a continued lack of progress in improving IT governance, determined that the CIO's visibility into investments across the agency continued to be limited, and identified flaws in the process developed to improve governance.[16] Specifically, the Office of Inspector General noted that the Office of the CIO had made changes to its IT governance boards over the past few years, but the boards had not made strategic decisions to substantively impact how NASA IT would be managed.

---

[14] GAO, Information Security: NASA Needs to Remedy Vulnerabilities in Key Networks, GAO-10-4 (Washington, D.C.: Oct. 15, 2009).

[15] National Aeronautics and Space Administration Office of Inspector General, NASA's Information Technology Governance, IG-13-015 (Washington, D.C.: Jun. 5, 2013).

[16] National Aeronautics and Space Administration Office of Inspector General, *NASA's Efforts to Improve the Agency's Information Technology Governance*, IG-18-002 (Washington, D.C.: Oct. 19, 2017).

According to the Office of Inspector General, slow implementation of the revised governance structure had left many IT officials operating under the previous inefficient and ineffective framework.

The report also noted that, as of August 2017, the Office of the CIO had not finalized the roles and responsibilities for IT management and lingering confusion regarding security roles, coupled with poor IT inventory practices, had negatively impacted NASA's security posture. Importantly, the report explained that the Office of the CIO continued to have limited influence over IT management within the mission directorates and at centers.

The Office of Inspector General made five recommendations to the CIO that were intended to improve, among other things, governance and security. As of October 2017, NASA had concurred with three recommendations, partially concurred with two recommendations, and described corrective actions taken or planned. However, the Office of Inspector General found that NASA's original proposed action to address the fourth recommendation was insufficient; thus, in December 2017, the agency established additional proposed actions to address that recommendation.

Key IT Management Disciplines

We have identified a set of essential and complementary management disciplines that provide a sound foundation for IT management. These include the following:

- *Strategic planning:* Strategic planning defines what an organization seeks to accomplish and identifies the strategies it will use to achieve desired results. We have previously reported that a defined strategic planning process allows an agency to clearly articulate its strategic direction and establish linkages among planning practices, such as goals, objectives, and strategies and identified leading practices for agency planning.[17]
- *Workforce planning:* We have previously reported that it is important for an agency to have a strong IT workforce to help ensure the timely and effective acquisition of IT.[18] In November 2016, we identified eight key workforce planning activities derived from the Clinger-Cohen Act of 1996 and relevant guidance,

[17] GAO, Social Security Administration: Improved Planning and Performance Measures Are Needed to Help Ensure Successful Technology Modernization, GAO-12-495 (Washington, D.C.: Apr. 26, 2012); Defense Business Transformation: Status of Department of Defense Efforts to Develop a Management Approach to Guide Business Transformation, GAO-09-272R (Washington, D.C.: Jan. 9, 2009); GAO-10-846G; and Library of Congress: Strong Leadership Needed to Address Serious Information Technology Management Weaknesses, GAO-15-315 (Washington, D.C.: Mar. 31, 2015).

[18] GAO, Standards for Internal Control in the Federal Government, GAO-14-704G (Washington, D.C.: Sept. 10, 2014); Human Capital: Key Principles for Effective Strategic Workforce Planning, GAO-04-39 (Washington, D.C.: Dec. 11, 2003); and A Model of Strategic Human Capital Management, GAO-02-373SP (Washington, D.C.: Mar. 15, 2002).

including memorandums and guidance from OPM and OMB, and prior GAO reports.[19] These laws and guidance focus on the importance of setting the strategic direction for workforce planning, analyzing the workforce to identify skill gaps, developing strategies to address skill gaps, and monitoring and reporting on progress in addressing skill gaps.

- *IT governance:* IT projects can significantly improve an organization's performance, but they can also become costly, risky, and unproductive. In 1996, Congress passed the Clinger-Cohen Act, which requires executive branch agencies to establish a process for selecting, managing, and evaluating investments in order to maximize the value and assess and manage the risks of IT acquisitions.[20] Agencies can maximize the value of their investments and minimize the risks of their acquisitions by having an effective and efficient governance process, as described in GAO's guide to effective IT investment management.[21]
- *Cybersecurity:* Federal systems and networks are often interconnected with other internal and external systems and networks, including the Internet. When systems are interconnected, the number of avenues of attack increases and the attack surface expands. Effective security for agency systems and data is essential to prevent data tampering, disruptions in critical operations, fraud, and inappropriate disclosure of sensitive information, including personal information entrusted to the government by members of the American public.[22] Taking action to assure that an agency's contractors and partners are adequately protecting the agency's information and systems is one way an agency can address cybersecurity risks.

  NIST has issued a suite of information security standards and guidelines that, collectively, provide comprehensive guidance on managing cybersecurity risk to agencies and any entities performing work on the agencies' behalf.[23] NIST's cybersecurity framework was issued in February 2014 in response to Executive

---

[19] GAO-17-8; GAO, Cybersecurity Human Capital: Initiatives Need Better Planning and Coordination, GAO-12-8 (Washington, D.C.: Nov. 29, 2011); Human Capital: Key Principles for Effective Strategic Workforce Planning, GAO-04-39 (Washington, D.C.: Dec. 11, 2003); and A Model of Strategic Human Capital Management, GAO-02-373SP (Washington, D.C.: Mar. 15, 2002).

[20] Clinger-Cohen Act of 1996, Pub. L. No. 104-106, Div. D and Div. E, § 5125(c)(3) 110 Stat. 642, 684-85 (Feb. 10, 1996), codified at 40 U.S.C. § 11315(c)(3).

[21] GAO-04-394G.

[22] GAO, Social Security Administration: Effective Planning and Management Practices Are Key to Overcoming IT Modernization Challenges, GAO-16-815T (Washington, D.C.: July 14, 2016).

[23] These documents include NIST, Managing Information Security Risk: Organization, Mission, and Information System View (Gaithersburg, Md.: March 2011); Security and Privacy Controls for Federal Information Systems and Organizations (Gaithersburg, Md.: April 2013), and the Framework for Improving Critical Infrastructure Cybersecurity (Gaithersburg, Md.: February 2014).

Order 13636.[24] The framework outlines a risk-based approach to managing cybersecurity risk and protecting an organization's critical information assets. Subsequent to the issuance of the cybersecurity framework, a May 2017 executive order required agencies to use the framework to manage cybersecurity risks.[25] The order outlined actions to enhance cybersecurity across federal agencies and critical infrastructure to improve the nation's cyber posture and capabilities against cybersecurity threats to digital and physical security.

NASA Has Not Yet Effectively Established and Implemented Leading IT Management Practices

NASA has not yet effectively established and implemented leading IT management practices for strategic planning, workforce planning, governance, and cybersecurity. Specifically,

- The agency's IT strategic planning process is not yet fully documented and its IT strategic plan lacks key elements called for by leading practices.
- NASA has not yet established an IT workforce planning process consistent with leading practices.
- The agency has taken recent action to improve its IT governance structure; however, it has not yet fully established that structure, documented improvements to its investment selection process, fully implemented investment oversight leading practices, or fully defined its policies and procedures for IT portfolio management.
- NASA has not fully established an effective approach to managing agency-wide cybersecurity risk. While it has designated a risk executive, the agency lacks a dedicated office to provide comprehensive executive oversight of risks. In addition, the agency-wide cybersecurity risk management strategy is currently in development, and the agency's information security program plan does not address all leading practices and has not been finalized. Further, policies and procedures for protecting NASA's information systems are in place, but the agency has not ensured that they are always current or integrated.

---

[24] Executive Order 13636, Improving Critical Infrastructure Cybersecurity, issued in February 2013, outlines an action plan for improving security for critical cyber infrastructure. This includes, among other things, requirements for NIST to develop a voluntary critical infrastructure cybersecurity framework and performance measures.

[25] Executive Order 13800, Strengthening the Cybersecurity of Federal Networks and Critical Infrastructure, 82 Fed. Reg. 22, 391, May 16, 2017.

NASA Has Not Fully Documented Its IT Strategic Planning Process

Leading practices of IT strategic planning established in OMB guidance call for an agency to document its IT strategic planning process, including, at a minimum, documenting the responsibilities and accountability for IT resources across the agency. It also calls for documenting the method by which the agency defines its IT needs and develops strategies, systems, and capabilities to meet those needs.[26]

NASA's documented IT strategic planning process describes the responsibilities and accountability for IT resources across the agency. For example, NASA has assigned specific governance bodies with responsibility for developing and overseeing the implementation of the IT strategy. Also, in its IT strategic plan, NASA described key stakeholders across the agency that are responsible for the development of the plan. These stakeholders include the Associate CIOs, representatives from mission directorates, mission support organizations, and the centers.

On the other hand, the methods by which the agency defines its IT needs and develops strategies, systems, and capabilities to meet those needs are not documented. For example, according to the IT strategic plan, the Office of the CIO is to perform a gap analysis to inform the development of NASA's roadmap that translates its IT needs and the strategies identified for meeting those needs into tactical plans. The tactical plans are to define how the strategic plan will be incrementally executed to achieve the longer term goals.

However, the Office of the CIO has not documented in its strategic planning policies and procedures how the CIO will perform the gap analysis or the methods for developing these tactical plans and roadmaps. This is particularly important since, according to officials in NASA's Office of the CIO, the centers vary as to whether they have developed their own IT strategic plans or tactical plans, and the office does not oversee or review any center-level plans to ensure they align with the NASA IT strategic plan.

According to officials in the Office of the CIO, NASA used a new model in formulating its IT strategy for fiscal years 2018 to 2021, such as including a broader set of stakeholders in the strategic planning cycle before documenting the strategic planning process. The officials stated that they intend to identify lessons learned from using this new model and formally document a complete and repeatable IT strategic planning process in the future. However, the agency has not established time frames for when the Office of the CIO will fully document its strategic planning process. Without a fully documented strategic

[26] OMB, Circular No. A-11: Preparation, Submission, and Execution of the Budget, July 2017; Circular No. A-130: Managing Information as a Strategic Resource, (Washington, D.C.: July 28, 2016) and Memorandum M-13-09 Fiscal Year 2013 PortfolioStat Guidance: Strengthening Federal IT Portfolio Management (Washington, D.C.: Mar. 27, 2013); GAO, Social Security Administration: Improved Planning and Performance Measures Are Needed to Help Ensure Successful Technology Modernization, GAO-12-495 (Washington, D.C.: Apr. 26, 2012); Defense Business Transformation: Status of Department of Defense Efforts to Develop a Management Approach to Guide Business Transformation, GAO-09-272R (Washington, D.C.: Jan. 9, 2009); GAO-10-846G; and Library of Congress: Strong Leadership Needed to Address Serious Information Technology Management Weaknesses, GAO-15-315 (Washington, D.C.: Mar. 31, 2015).

planning process, NASA risks not being able to clearly articulate what it seeks to accomplish and identify the IT resources needed to achieve desired results in a way that is consistent and complete.

NASA Has Improved Its IT Strategic Plan, but Has Not Yet Established a Comprehensive Plan

In addition to calling for agencies to fully document the strategic planning process, leading practices from OMB guidance and our prior research and experience at federal agencies have shown that an agency should develop a comprehensive and effective IT strategic plan that (1) is aligned with the agency's overall strategy; (2) identifies the mission of the agency, results-oriented goals, and performance measures that permit the agency to determine whether implementation of the plan is succeeding; (3) includes strategies, with resources and time frames, that the governing IT organization intends to use to achieve desired results; and (4) provides descriptions of interdependencies within and across projects so that they can be understood and managed.[27] The resulting plan is to serve as an agency's vision, or road map, and help align information resources with business strategies and investment decisions.

NASA has taken steps to improve its IT strategic plan, but the updated plan is not comprehensive in that it does not fully address all four elements of a comprehensive and effective plan outlined above. In this regard, the agency had a prior strategic plan covering the time frame of March 2014 to November 2017. More recently, in December 2017, the CIO and Associate Administrator approved an updated plan for implementation. The updated plan is intended for use from the date it was approved through fiscal year 2021.

Regarding the four elements of a comprehensive IT strategic plan, NASA's prior plan addressed one element, partially addressed two elements, and did not address one element. The updated plan was slightly improved in that it addressed two elements, partially met one element, and did not meet one element of a comprehensive strategic plan. Table 1 provides a summary of the extent to which NASA's prior IT strategic plan (covering the time frame of March 2014 to November 2017) and recently updated IT strategic plan (covering the time frame of December 2017 to fiscal year 2021) addressed key elements of a comprehensive strategic plan.

NASA's prior IT strategic plan was aligned with the agency's overall strategic plan and identified the mission of the agency and results-oriented goals. However, these goals

---

[27] OMB, Circular No. A-11: Preparation, Submission, and Execution of the Budget, July 2017; Circular No. A-130: Managing Information as a Strategic Resource, (Washington, D.C.: July 28, 2016) and Memorandum M-13-09 Fiscal Year 2013 PortfolioStat Guidance: Strengthening Federal IT Portfolio Management (Washington, D.C.: Mar. 27, 2013); GAO, Social Security Administration: Improved Planning and Performance Measures Are Needed to Help Ensure Successful Technology Modernization, GAO-12-495 (Washington, D.C.: Apr. 26, 2012); Defense Business Transformation: Status of Department of Defense Efforts to Develop a Management Approach to Guide Business Transformation, GAO-09-272R (Washington, D.C.: Jan. 9, 2009); GAO-10-846G; and Library of Congress: Strong Leadership Needed to Address Serious Information Technology Management Weaknesses, GAO-15-315 (Washington, D.C.: Mar. 31, 2015).

were not linked to specific performance measures that were needed to track progress and did not always describe strategies to achieve desired results. Additionally, this plan lacked descriptions of interdependencies within and across projects.

NASA's updated IT strategic plan is aligned with the agency's overall strategic plan and identifies the mission of the agency and results-oriented goals.[28] For example, the plan describes the agency's IT vision, mission, principles, and objectives of five strategic goals—excellence, data, cybersecurity, value, and people. To support these goals, the plan defines 14 objectives to be accomplished over 4 years. For example, the plan defines objectives for increasing the effectiveness of NASA's IT strategy execution through disciplined program and project management.

Table 1. Extent to Which the National Aeronautics and Space Administration's (NASA) Prior and Updated Information Technology (IT) Strategic Plans Addressed Key Elements of a Comprehensive Strategic Plan

Key element	GAO assessment of NASA's prior IT strategic plan	GAO assessment of NASA's updated IT strategic plan
Aligns with the agency's overall strategy	NASA's IT strategic plan addressed the key element	NASA's IT strategic plan addressed the key element
Identifies the mission of the agency, results-oriented goals, and performance measures	NASA's IT strategic plan partially addressed the key element	NASA's IT strategic plan addressed the key element
Includes strategies to achieve desired results	NASA's IT strategic plan partially addressed the key element	NASA's IT strategic plan partially addressed the key element
Provides descriptions of interdependencies among projects	NASA's IT strategic plan did not address the key element	NASA's IT strategic plan did not address the key element

Source: GAO analysis of National Aeronautics and Space Administration (NASA) data. | GAO-18-337.

In addition, NASA has improved upon the prior plan by identifying performance measures that allow the agency to determine whether it is succeeding in the implementation of its goals. For example, in order to increase the effectiveness of its IT strategy execution, the Office of the CIO expects 85 percent of projects to be in conformance with approved project plans by the end of fiscal year 2018. As another example, to prepare its employees to achieve NASA's IT vision, the Office of the CIO plans to, by the end of fiscal year 2020, identify skills gaps and ways to close the gaps based on the workforce strategy.

However, similar to the prior plan, the updated plan does not fully describe strategies NASA intends to use to achieve the desired results or descriptions of interdependencies

[28] NASA's 2018 agency-wide strategic plan was finalized on February 12, 2018.

within and across projects. Specifically, the plan discusses how the agency intends to achieve its strategic goals and objectives through various activities. For example, according to the plan, to increase the effectiveness of investment analysis and prioritization, NASA intends to implement a financial management process that integrates Office of the CIO, center, and mission directorate IT spending. The plan states that this process will map IT investments to NASA's vision and strategy, as well as enable high-quality internal and external investment insight and reporting.

However, the updated plan does not further describe the strategies NASA intends to use to accomplish these activities, including a schedule for significant actions and the resources needed to achieve this objective. For instance, the plan states that the Office of the CIO will define clear lines of authority and accountability for IT between the agency and NASA's centers, but does not describe a strategy, including time frames and resources, for accomplishing this. Additionally, the plan does not describe interdependencies between projects, which is essential to help define the relationships within and across projects and major initiatives.

According to NASA's CIO, the updated strategic plan was kept at a higher level with the expectation that more detailed implementation plans (e.g., tactical plans and roadmaps) would define the necessary projects and interdependencies. However, NASA has not defined guidance for developing the implementation plans to ensure that any plans developed will fully describe strategies and interdependencies, or time frames for when these plans will be completed. Until NASA incorporates the key elements of a comprehensive IT strategic plan, it will lack critical information needed to align information resources with business strategies and investment decisions.

NASA Has Gaps in Its IT Workforce Planning Efforts

Key to an agency's success in managing its IT investments is sustaining a workforce with the necessary knowledge, skills, and abilities to execute a range of management functions that support the agency's mission and goals. Achieving such a workforce depends on having effective human capital management consistent with workforce planning activities pursuant to federal laws[29] and guidance.[30]

---

[29] Federal Cybersecurity Workforce Assessment Act of 2015, Pub. L. No. 114-113, Div. N, Title III, 129 Stat. 2242, 2975-77 (Dec. 18, 2015); Carl Levin and Howard P. "Buck" McKeon National Defense Authorization Act for Fiscal Year 2015, Div. A, Title VIII, Subtitle D—Federal Information Technology Acquisition Reform, Pub. L. No. 113-291, § 835, 128 Stat. 3292, 3449 (Dec. 19, 2014), codified at 41 U.S.C. § 1704 note; E-Government Act of 2002, Pub. L. No. 107-347, § 209 (Dec. 17, 2002), 44 U.S.C. § 3501 note; and Pub. L. No. 104-106, Div. D and Div. E, § 5125(c)(3) (Feb. 10, 1996), codified at 40 U.S.C. § 11315(c)(3).

[30] In 2002, OPM released a *Human Capital Assessment and Accountability Framework*— developed jointly with GAO and OMB—that identifies five human capital systems that together provide a consistent, comprehensive representation of human capital management for the federal government (http://www.opm.gov/hcaaf_resource_center/, accessed July 1, 2016). In addition, GAO, OPM, and OMB have established subsequent

Specifically, OMB requires agencies to develop and maintain a current workforce planning process.[31] In addition, we reported in 2016 on the importance of setting a strategic direction for IT workforce planning, identifying skills gaps and implementing strategies to address them, and monitoring and reporting on progress in addressing the identified skills gaps.[32] We identified eight key IT workforce planning activities that are essential to agency efforts to establish an effective IT workforce:

1. establish and maintain a workforce planning process;
2. develop competency and staffing requirements;
3. assess competency and staffing needs regularly;
4. assess gaps in competencies and staffing;
5. develop strategies and plans to address gaps in competencies and staffing;
6. implement activities that address gaps (including IT acquisition cadres, cross-functional training of acquisition and program personnel, career paths for program managers, plans to strengthen program management, and use of special hiring authorities);
7. monitor the agency's progress in addressing competency and staffing gaps; and
8. report to agency leadership on progress in addressing competency and staffing gaps. [33]

The Office of the CIO has had IT workforce planning efforts underway since 2015 that are intended to address the workforce planning activities listed above; however, the office has not finalized or implemented any of the planned actions. The office recently began working to establish a more comprehensive workforce strategy for fiscal year 2019 to align with the agency's increased emphasis on improving the overall workforce. Specifically, in the draft NASA Strategic Plan, the agency established a workforce development goal and two strategic objectives that relate to its IT workforce and call for, among other things, workforce training and efforts to increase cybersecurity awareness to reduce cybersecurity risks.

---

guidance on key principles and steps associated with workforce planning that agencies can utilize in their efforts to assess and address IT skill gaps. GAO-17-8 and OPM, *Workforce Planning Model*, https://www.opm.gov/policy-data-oversight/ human-capital-management/reference-materials/ (accessed June 7, 2016). OPM, *IT Program Management Career Path Guide*, Nov. 18, 2011 and, for example, OMB, Circular A-130, *Managing Information as a Strategic Resource* (Washington, D.C.: July 27, 2016); *Federal Cybersecurity Workforce Strategy*, Memorandum M-16-15 (Washington, D.C.: July 12, 2016); *Cybersecurity Strategy and Implementation Plan (CSIP) for the Federal Civilian Government*, Memorandum M-16-04 (Washington, D.C.: Oct. 30, 2015); *Management and Oversight of Federal Information Technology*, Memorandum M-15-14 (Washington, D.C.: June 10, 2015); and *Guidance for Specialized Information Technology Acquisition Cadres* (Washington, D.C.: July 13, 2011); and *25 Point Implementation Plan to Reform Federal Information Technology Management* (Washington, D.C.: Dec. 9, 2010).

[31] OMB, Circular A-130, Managing Information as a Strategic Resource (Washington, D.C.: July 27, 2016).

[32] GAO-17-8; and GAO, Human Capital: Key Principles for Effective Strategic Workforce Planning, GAO-04-39 (Washington, D.C.: Dec. 11, 2003).

[33] GAO-17-8 and GAO-04-39.

Nevertheless, NASA has gaps in its IT workforce planning efforts. Of the eight key IT workforce planning activities that we previously outlined, NASA partially implemented five and did not implement three. Table 2 shows the extent to which NASA has implemented each IT workforce planning activity and provides examples of workforce practices planned or implemented, as well as those not yet undertaken.

Table 2. Evaluation of the National Aeronautics and Space Administration's (NASA) Implementation of Key Information Technology (IT) Workforce Planning Activities

Key activities	Rating	Examples of NASA's efforts to address relevant practices
Establish and maintain a workforce planning process	Partially implemented	NASA's Chief Information Officer (CIO) has taken steps to establish elements of a workforce planning process. Specifically, the CIO is responsible for IT workforce planning and requires center CIOs to implement workforce processes, hire appropriate IT skill sets, and offer training and development opportunities. In addition, the agency has established elements of the workforce planning process for how to respond to changing mission priorities and IT. Further, through its annual budget guidance, NASA's Office of Human Capital Management provides maximum annual staffing limits and offers strategic planning and budget development guidance for workforce planning, and its Office of the CIO identifies recruiting and training activities for high-priority positions and for retaining staff through IT professional career tracks. However, the process is not complete because NASA lacks procedures for how the process will be implemented. For example, NASA's desk guide for workforce planning provides only a general description of workforce practices and lacks details about how the practices are to be implemented. NASA also has not established how it will maintain a current IT workforce planning process. Further, the agency has not yet finalized or implemented a human capital operating plan.
Develop competency and staffing requirements	Partially implemented	To address this key activity, the Office of the CIO identified 15 mission-critical IT workforce competencies and skills. The Office of the CIO also assessed staffing requirements by reviewing staffing for business IT, analyzing mission staffing data submitted by centers, and identifying current and future IT staffing estimates. However, the agency has not formally established competencies or staffing requirements in an approved IT workforce plan that could be linked directly to strategic and annual performance plans.
Assess competency and staffing needs regularly	Not implemented	NASA did not implement this key activity. According to agency officials, including the Associate CIO for Capital Planning and Governance, the Office of the CIO does not yet assess competencies or staffing needs regularly. Further, we determined that the agency has not projected staffing needs far enough into the future (i.e., 3 to 5 years) to ensure that estimates remain aligned with the agency's long-term goals and objectives.

Table 2. (Continued)

Key activities	Rating	Examples of NASA's efforts to address relevant practices
Assess gaps in competencies and staffing	Partially implemented	While the agency has taken action to assess gaps in competencies and staffing, the IT workforce gap assessments are outdated and not regularly updated. NASA's CIO is responsible for working with the Office of Human Capital Management to analyze gaps in competencies and staffing, and an assessment was performed in 2015. However, NASA has not since updated its assessment and its policy does not require the CIO to conduct gap analyses on a regular basis. According to the agency's new IT strategic plan, the Office of the CIO plans to identify skills gaps based upon a new workforce strategy by fiscal year 2020.
Develop strategies and plans to address gaps in competencies and staffing	Partially implemented	NASA's CIO has taken specific actions to address this key activity. Specifically, the CIO has delegated certain tasks associated with strategies and plans for addressing gaps in competencies and staffing to staff within the Office of the CIO or center CIOs. In so doing, the CIO and her designees identify, recruit, hire, train, and provide annual performance reviews for officials leading major IT programs. In addition, staffing and skills requirements are identified via program reviews. Staff for major IT programs are recruited and hired through NASA's recruiting and hiring process. Annual performance reviews are conducted and training needs are assessed between the direct supervisor and the employee. Center CIOs are responsible for training and expertise of enterprise programs assigned to that center. However, NASA has not yet fully developed strategies or plans to address identified gaps. Specifically, NASA has not yet established a talent management plan to close staffing and competency gaps, manage staffing surpluses, maintain the strengths of the existing workforce, or mitigate risks. The chief of the Office of Human Capital Management reported that the agency is currently developing a talent management program intended to consolidate competencies, staffing requirements, talent management, and recruitment; however, the program is not expected to be complete until fiscal year 2019.
Implement activities that address gaps (including IT acquisition cadres, cross-functional training of acquisition and program personnel, career paths for program managers, plans to strengthen program management, and use of special hiring authorities)	Not implemented	NASA has not yet implemented activities to address gaps identified. NASA's CIO and other officials from the Office of the CIO reported that they have not developed the required acquisition human capital plan. As a result, they have not yet performed the critical prerequisites for the plan, including analyzing current IT acquisition staffing challenges; determining if developing or expanding the use of cadres would improve IT program results; outlining a plan to pilot or expand cadres for an especially high-risk IT area; or addressing how the agency is meeting its human capital requirements to support timely and effective acquisitions.

Key activities	Rating	Examples of NASA's efforts to address relevant practices
Monitor the agency's progress in addressing competency and staffing gaps	Partially implemented	NASA has begun to address this key activity by developing policies calling for the agency to monitor progress in addressing competency and staffing gaps and requiring the CIO to approve the selection of center CIOs and provide input into their performance evaluations. Specifically, NASA Policy Document 2800.1B – Managing Information Technology directs the CIO, in coordination with the Office of Human Capital Management, to regularly conduct a gap analysis and develop and execute a strategy for matching NASA's needs with the required IT workforce skills. However, NASA has not established procedures for how progress will be measured or required the CIO to assess progress annually. The agency also developed criteria it planned to use to determine if efforts to monitor progress had been successful but never incorporated the criteria in approved plans. Officials in the Office of the CIO stated that NASA policy is currently being updated to add a requirement to annually conduct these key activities, but time frames for when this will be completed have not been established.
Report to agency leadership on progress in addressing competency and staffing gaps	Not implemented	NASA has not implemented this activity. The Office of the CIO does not yet report to agency leadership on progress in addressing competency and staffing gaps. According to the Associate CIO for Technology and Innovation, reports are not made to agency leadership on IT workforce gaps. Officials in the Office of the CIO stated that NASA intends to update its policy in the future to add an annual requirement to report to head of agency on progress made in improving IT personnel.

Legend: Partially implemented - NASA's IT workforce policies, procedures and planning documents addressed some, but not all of the practices for the leading activity.

Not implemented - NASA's IT workforce policies, procedures and planning documents did not address any of the practices for the leading activity.

Source: GAO analysis of National Aeronautics and Space Administration (NASA) data. | GAO-18-337.

According to NASA's CIO, the Office of the CIO put IT workforce planning activities on hold in 2015 pending the outcome of more comprehensive, agency-wide efforts. Specifically, the agency began planning and developing a new phased program—the Mission Support Future Architecture Program—designed to deliver workforce and other mission support services, including a talent management program.[34] Phase 1 of the new phased Mission Support Future Architecture Program began in May 2017.

According to the NASA CIO, the Office of the CIO is expected to be part of a future phase and to renew its IT workforce planning as part of that effort. However, the CIO did not have an estimate for when the Office of the CIO would join the program. Until NASA implements all of the key IT workforce planning activities discussed in this report, the agency will have difficulty anticipating and responding to changing staffing needs. Further,

[34] NASA's Executive Council took action in May 2017 to improve the agency's efficiency and effectiveness by establishing the Mission Support Future Architecture Program for, among other things, workforce planning. The program engaged a team to develop an overall implementation plan, a management structure, and an approach for executing a phased enterprise architecture for mission support services.

NASA will face challenges in controlling human capital risks when developing, implementing, and operating IT systems.

NASA's IT Governance Approach Does Not Fully Address Leading Practices

Leading practices for governing IT, such as those identified by GAO in its IT investment management framework, call for agencies to establish and follow a systematic and organized approach to investment management to help lay a foundation for successful, predictable, and repeatable decisions.[35] Critical elements of such an approach include instituting an IT investment board (or boards), developing and documenting a governance process for investment selection and for investment oversight, and establishing governance policies and procedures for managing the agency's overall IT investment portfolio.

NASA Has Not Fully Instituted an Effective Governance Structure

Instituting an effective IT governance structure involves establishing one or more governance boards, clearly defining the boards' roles and responsibilities, and ensuring that they operate as intended. Moreover, Section 811(a) of the National Aeronautics and Space Administration Transition Authorization Act of 2017 directs the agency to ensure that the NASA CIO, mission directorates, and centers have appropriate roles in governance processes. The act also calls on the Administrator to provide, among other things, an IT program management framework to increase the efficiency and effectiveness of IT investments, including relying on metrics for identifying and reducing potential duplication, waste, and cost.

Table 3. National Aeronautics and Space Administration (NASA) Information Technology (IT) Governance Councils and Boards

Council or board	Role in IT governance	Minimum frequency of meetings	Thresholds	Member(s)
Mission Support Council (agencywide body with responsibility beyond IT)	Senior NASA decision-making body responsible for managing mission support issues, including those that require a high degree of integration or are highly visible. Its scope of authority includes the agency's mission support investments in facilities,	Monthly	Not applicable	• Deputy Associate Administrator (chair) • Associate Administrator • Associate Administrator for Mission Support • Chief Information Officer (CIO) • Chief Financial Officer • Chief, Safety and Mission

[35] GAO-04-394G.

Council or board	Role in IT governance	Minimum frequency of meetings	Thresholds	Member(s)
	workforce, infrastructure, and IT. For IT, this council is to review the IT strategy, enterprise architecture, and IT policy changes. The council chartered the IT Council to govern IT. When the IT Council cannot resolve governance decisions, they are escalated to this council.			
Agency Program Management Council (agencywide body with responsibility beyond IT)	Senior decision-making body that is responsible for baselining and assessing the performance of NASA projects, programs, and investments by mission directorates and across the agency. It is intended to ensure successful outcomes supporting the achievement of NASA strategic goals. This council is responsible for approving highly-specialized IT, regardless of the size of the investment.	Monthly	Not applicable	• Associate Administrator (Chair) • Deputy Associate Administrator • Chief Engineer • Chief, Safety and Mission Assurance • Associate Administrators for the Directorates • CIO • Chief Financial Officer • General Counsel • Chief Health and Medical Officer • Chief Scientist • Chief Technologist
IT Council	IT-specific governance board and also NASA's executivelevel IT governance board. According to its charter, this council sets policy for all of NASA's IT, including highlyspecialized IT. However, its program management authority is limited to IT services delivered and managed by the CIO (i.e., IT that is not highly-specialized).	Monthly[a]	Above $10 million	• CIO (chair) • Senior executive leadership from mission directorates • Senior executive leadership from field centers • Senior executive leadership from NASA • mission support offices

Table 3. (Continued)

Council or board	Role in IT governance	Minimum frequency of meetings	Thresholds	Member(s)
CIO Leadership Team	Advisory board for the CIO intended to serve as a change agent and sounding board providing visibility into the IT requirements, operations, performance, risk management strategies, and stakeholder issues for the centers and mission directorates. According to its charter, the team's responsibilities include (1) overseeing the implementation of agency IT strategy and policy; (2) identifying opportunities and investment recommendations; (3) assessing the impact of and providing recommendations on proposed center and mission directorate investments; and (4) reviewing or recommending decisions before they are submitted to the IT Council.	Every other week[b]	$1-10 million	• Deputy CIO (chair—formally delegated by NASA CIO) • Associate CIO for Capital Planning and Governance • Associate CIO for IT Security • Associate CIO for Technology and Innovation • Associate CIO for Enterprise Services and Integration • Senior Advisor, Cybersecurity • A representative from NASA's Shared Services Center • Nine field center CIOs as well as the CIO from the Jet Propulsion Laboratory • IT representatives from NASA's four mission directorates
IT Program Management Board	This board is to provide a forum for high-level agency participation in the oversight and evaluation of NASA's IT programs and projects. Specifically, the board is to oversee IT programs and projects from development through implementation. It is to conduct key decision point reviews to ensure that programs and projects meet	Bimonthly	Less than $1 million	• Deputy CIO (chair—formally delegated by NASA CIO) • Division Chief, Office of the CIO Capital Planning and Governance Division • Representative from the CIO Leadership Team • Representative from the Office of the Chief Engineer • Center representative (two) • Mission support directorate representative (two) • Enterprise Architecture Lead (non-voting member) • · IT Security representative (non-voting member)

Council or board	Role in IT governance	Minimum frequency of meetings	Thresholds	Member(s)
	their cost, schedule, and scope commitments.			•
Center-specific IT governance boards	Responsible for approving center-specific IT investments.	Varies by center[c]	Less than $1 million	• Center CIOs (chair) • At the Goddard Space Flight Center, the tactical planning group responsible for IT investment management consists of staff appointed by the directors of the center's directorates • Johnson Space Center's governance council also includes the representatives from the center's directorates, but it also includes a center Chief Financial Officer representative and representatives from the procurement office and certain mission programs

Source: GAO analysis of National Aeronautics and Space Administration (NASA) information. | GAO-18-337.

[a] After the IT Council began meeting, members decided to meet monthly instead of quarterly and updated the charter in February 2018 to reflect this change.

[b] As of April 2018, NASA's IT governance lead stated that the agency intended to update the team's charter to call for more frequent meetings. Our review of documentation provided by NASA showed that the agency had scheduled the team to meet during 2018 on every week except for the one week per month when the IT Council is scheduled to meet.

[c] The NASA documentation we reviewed did not identify an established minimum frequency for center board meetings. As such, if NASA centers establish center-specific IT governance boards, the centers are to determine how frequently each board should meet.

NASA has established three boards focused specifically on IT governance—an IT Council which is its executive-level IT board, a CIO Leadership Team, and an IT Program Management Board which provides oversight of programs and projects. Meeting minutes for the three IT-specific governance bodies identified above revealed that these groups are meeting as required by their charters.

Further, two of NASA's agency-wide councils (whose governance responsibilities extend beyond IT) also play a role in IT governance. Specifically, the Mission Support Council is the governance body to which the IT Council escalates unresolved decisions, and the Agency Program Management Council is responsible for reviewing and approving highly-specialized IT. In addition, NASA centers have the option to create center-specific IT governance boards to make decisions about center-level IT investments under the authority of center CIOs.

Table 3 describes the roles of the IT-specific governance boards, the agency-wide councils with roles in IT governance, and the center-level IT governance boards. The table also includes additional details on how frequently the councils and boards meet, the dollar thresholds NASA has established to determine which investments each council or board reviews, and which officials serve as members of the boards.

Although it has established and assigned responsibilities for the aforementioned governance councils and boards, NASA has not yet fully instituted an effective investment board governance structure for several reasons.

- *Planned improvements to the IT governance structure are not yet complete.* NASA has established new governance boards in addition to the boards listed above, but has not yet approved charters to guide their operations. Specifically, the Office of the CIO has revised its governance structure to establish six new boards, one for each of its IT programs. Agency officials, including the IT governance lead, reported that the boards had been established; however, as of December 2017, NASA had not yet approved charters defining the new governance bodies' membership, functions, and interactions with other governance boards.[36]
- *Roles and responsibilities of the IT governance boards and agency-wide governance councils are not clearly defined.* NASA continues to operate a federated governance model with decentralized roles and responsibilities for governance of mission and business IT investments. Business IT is selected and approved by the IT-specific governance boards, but mission IT follows a different path for investment selection in that it is not reviewed and approved by the CIO along with other IT investments proposed for selection. Instead, the Agency Program Management Council's reviews focus on the selection of overall mission programs, and not on selecting IT. As a result, mission IT has historically been reported to the Office of the CIO only if the program has been designated as a major agency IT investment to be reported to OMB.

  NASA has begun making changes to its decentralized governance approach in response to provisions in legislation commonly referred to as the Federal Information Technology Acquisition Reform Act[37] that are intended to ensure that the CIO has visibility into both mission and business IT investments. However, the agency has not yet developed policies and procedures to clarify how these changes will affect the CIO's and governance boards' roles and responsibilities. For example, in January 2017, the IT Council approved an updated definition for

---

[36] The NASA Inspector General reported in October 2017 that the program boards had not been implemented. See National Aeronautics and Space Administration Office of Inspector General, *NASA's Efforts to Improve the Agency's Information Technology Governance*, IG-18-002, (Washington, D.C.: Oct. 19, 2017).

[37] Federal Information Technology Acquisition Reform provisions of the Carl Levin and Howard P. 'Buck' McKeon National Defense Authorization Act for Fiscal Year 2015, Pub. L. No. 113-291, div. A, title VIII, subtitle D, § 831 (Dec. 19, 2014).

highly-specialized IT[38] and established new expectations about the extent to which highly-specialized IT investments would be reviewed by the NASA CIO. However, NASA has not clarified roles and responsibilities for identifying such investments and ensuring they are reported by mission directorate programs to the CIO. In addition, the agency has not yet outlined procedures for how these investments that are overseen by the agency-wide Agency Program Management Council are to be reported to the CIO or IT-specific governance boards.

During a January 2017 IT Council meeting, the NASA CIO acknowledged that roles and responsibilities for IT governance were unclear and that it would take 1 to 2 years to clarify them. In July 2017, the Deputy CIO recognized that significant work remained for NASA to achieve a consistent agency-wide governance approach with established roles and responsibilities.

- *While the IT governance boards are meeting regularly, they are not consistently operating as intended.* Board charters finalized in 2016 defined the membership for the governance boards and established expectations for the expertise to be made available to support board decisions. However, the boards are not consistently operating with all designated board members in attendance. For example, the Chief Engineer was designated as a member of the IT Council, but the council's meeting minutes indicated that the Deputy Chief Engineer regularly attends the council meetings instead.
- In addition, IT Program Management Board meetings are consistently held with fewer voting members than designated by the board's charter. The board's meeting minutes indicated that fewer than six voting members regularly attend board meetings instead of the eight voting members outlined in the board charter. For example, the minutes showed that each meeting has been held with only one center and mission support directorate representative—instead of the two required by the charter.

NASA officials, including the Associate CIO for Capital Planning and Governance, stated that planned efforts to update the governance structure and develop additional guidance for IT investment management have impacted the agency's time frames for fully establishing its new boards and defining their roles and responsibilities. Specifically, these officials stated that the Office of the CIO is working to develop a comprehensive IT framework intended to update the governance structure, fully establish the new governance boards, and define governance roles and responsibilities. According to the officials, this framework is expected to be finalized in 2018, but the office did not provide a detailed schedule with milestones for completing the framework. Without a detailed schedule for

---

[38] NASA clarified that highly-specialized IT is "any equipment, system and/or software that is used in the acquisition, storage, retrieval, manipulation and/or transmission of data or information which comprises or is embedded in a mission platform, or a platform required for mission simulation, execution or operations."

updating the governance structure and establishing a comprehensive IT framework to help ensure that the revised governance boards are fully established and operating as intended, NASA may not be able to improve IT governance in accordance with the requirements in the National Aeronautics and Space Administration Transition Authorization Act of 2017.

NASA Has Not Completed or Updated Governance Selection Process Policies and Procedures and Lacks Established Guidance for Reselecting Investments

According to our IT investment management guide, defining policies and procedures for selecting investments provides investment boards and others with a structured process and a common understanding of how investments will be selected. Selection policies and procedures should, among other things, establish thresholds or criteria (e.g., investment size, technical difficulty, risk, business impact, customer needs, and cost-benefit analysis) for boards to use in identifying, analyzing, prioritizing and selecting new IT proposals.

In addition, outlining a process for reselecting ongoing projects is intended to support board decisions about whether to continue to fund projects not meeting established goals or plans. Using the defined selection process promotes consistency and transparency in IT governance decision making. Further, after the guidance has been developed, organizations must actively maintain it, making sure that it always reflects the board's current structure and the processes that are being used to manage the selection of the organization's IT investments.

NASA's defined selection process policies and procedures designated the CIO with responsibility to ensure that IT governance, investment management, and program/project management processes are integrated to facilitate the selection of appropriate IT investments. The agency has established multiple policies and procedures outlining certain aspects of how both mission programs and business IT investments are to be planned, such as standardized templates for requesting approval to plan investments and direction for teams to use in planning for investments. In addition, the Office of the CIO has established a *Capital Planning and Investment Control Guide* for business IT investments and issues annual budget guidance for requesting funding for IT investments.

The agency's selection process also includes specific IT governance processes developed by centers for the investments they review. For example, Goddard Space Flight Center had developed additional center-specific guidance assigning lead responsibility for assessing new and ongoing projects. The center also has established predetermined criteria, such as whether projects conflict, overlap, or are redundant with other projects, and the risk if the investment was not funded.

Nevertheless, NASA's established process does not yet define thresholds or criteria (e.g., qualitative or quantitative data) to be analyzed and compared when governance boards make decisions to select investments. Charters for NASA's governance boards outline the functions these boards are to perform and direct them to be involved in IT governance. However, the charters do not outline specific thresholds or procedures that the boards are to follow in selecting investments.

For example, NASA's process does not fully define how investment risks are to be evaluated. NASA policy establishes dollar thresholds for IT governance board reviews, but does not define any other parameters for how risk will be evaluated. In addition, NASA has established an expectation that the new capital investment review process is to yield risk-based decisions for all investments and help mitigate IT security risks. However, guidance for capital investment reviews does not address how investment risks are to be evaluated.

Moreover, NASA's selection process policies and procedures have not been updated to reflect efforts to improve governance. Its guidance for selecting investments (and for all aspects of its governance process) is fragmented, and the agency has not updated its policies and procedures to reflect current selection practices. In addition, this guidance does not yet reflect recent efforts to clarify and standardize the definitions of fundamental IT investment terms, such as "information technology" and "major" investments.

Further, while NASA has begun changing its selection process to ensure that the CIO and IT governance boards will be provided data about all IT investments, including mission IT investments such as highly-specialized IT, the agency's selection policies have not been updated to reflect these changes. NASA's *Capital Planning and Investment Control Guide* does not require all investments to be included in the selection process (or other IT governance processes) and the NASA *Space Flight Program and Project Management* procedures for mission program governance do not address whether or how the investments within mission programs are to be reported to the agency's IT-specific governance boards.

In addition, NASA has not yet defined a reselection process for IT investments. Current policies and guidance for selecting investments do not clearly define a consistent approach for how performance is to be considered in reselecting investments. Without a defined reselection process, the agency's boards lack structure and a common understanding about how to make decisions about whether to continue to fund projects not meeting established goals or plans.

NASA officials acknowledged that the current policies and procedures do not establish sufficient content within the business cases and IT plans for proposed investments to support effective governance decision making. The agency has begun working to update its policy for IT program and project management but did not expect to complete the update until April 2018. Further, even when this key IT investment management policy is updated, the agency will still need to update related policies and procedures to reflect changes it has made but not yet documented in the investment selection process. NASA has not yet established plans for when all needed updates to the policies and procedures will be completed.

Until NASA updates its IT governance policies and procedures to establish thresholds and procedures to guide its boards in decision making and outline a process for reselecting investments, the agency will be limited in its assurance that the investment selection process will provide a consistent and structured method for selecting investments. Further, until all relevant governance policies and procedures are updated to reflect current investment selection practices and proposed changes intended to provide the CIO with data about mission IT, the CIO will not be positioned to minimize investments that present undue risk to the agency and ensure accountability for both business and mission IT.

NASA Lacks Criteria for Assessing Investment Performance and Ensuring Oversight of All Investments

Organizations that provide effective IT investment oversight have documented policies and procedures that, among other things, ensure that data on actual performance (e.g., cost, schedule, benefit, and risk) are provided to the appropriate IT investment board(s). In addition, such organizations establish procedures for escalating or elevating unresolved or significant issues; ensure that appropriate actions are taken to correct or terminate underperforming IT projects based on defined criteria; and regularly track corrective actions until they are completed.

As with investment selection, NASA has established multiple policies and procedures for the oversight of IT investments. In October 2015, the agency added to its oversight processes by establishing a capital investment review process to improve the quality of the information available for investment oversight and established a matrix defining dollar thresholds to delineate oversight among the IT governance boards. The IT Program Management Board is also assigned specific oversight responsibilities for reviewing investment cost, schedule, performance, and risk at key lifecycle decision points for investments submitted for its review. In addition, the IT Program Management Board's charter requires this board to track, among other things, board decisions about investments and action items.

In implementing NASA's oversight practices, the IT Program Management Board consistently reviewed updates on investment performance (i.e., cost, schedule, and benefits) and progress. In addition, the IT Program Management Board's oversight decisions about IT investments are documented in meeting minutes,[39] and the board also records any action items identified for investments in the decision memorandums it submits to the CIO.

Nevertheless, we identified limitations in NASA's established oversight policies and procedures. For example, the agency's policies and procedures require IT investments to report data to the governance boards at key decision points but do not establish specific thresholds or other criteria for the governance boards to use in overseeing the investments' performance or escalating investments to review by other boards. The oversight guidance also does not specify the conditions under which a project would be terminated.

---

[39] Johnson Space Center standardized documentation for all center IT governance decisions in meeting minutes and consistently recorded additional supporting detail about IT decisions.

In addition, weaknesses we identified in oversight of specific NASA IT investments highlighted additional limitations of the established oversight process.

- Specifically, NASA did not have a mechanism for alerting the IT Program Management Board to provide oversight if investments were underperforming or overdue for review. For example, significant schedule overruns did not trigger additional oversight for one investment. In March 2015, NASA approved the proposed design for an investment to implement a security tool in June 2015 at an expected cost of $1.3 million. Although the project fell 13 months behind schedule and encountered unforeseen challenges, the IT Program Management Board did not review the investment again until June 2017—2 years later.[40]
- Not all IT investments followed the established oversight process. For example, in our review of governance board meeting minutes and documentation, we identified an investment that was close to completion before the IT Program Management Board reviewed its proposed design. Specifically, in February 2016, the board was asked—1 day before the investment was to become operational—to (1) approve the proposed design and (2) grant authority to operate for the investment intended for use by NASA staff and external partners. Although concerns about limited oversight were noted, the investment was approved.[41]
- Further, NASA lacks procedures to ensure that action items identified are tracked. We identified instances in which the IT Program Management Board did not consistently track action items identified for IT investments. NASA's investments typically report back to the IT Program Management Board at future decision point reviews about steps taken to address documented action items. However, the board's meeting minutes and documentation identified multiple examples of investments that were returned to the board at future decision points without reporting on whether identified action items had been addressed.

Moreover, NASA's oversight processes do not encompass highly-specialized or other IT that supports mission programs. After reviewing NASA's fiscal year 2015 budget request, OMB directed NASA to identify unreported IT investments throughout the agency to ensure that all related spending would be documented.[42] NASA established a team in

---

[40] The CIO was informed only that the investment had been approved; the memorandum did not document the significant concerns of governance board members or the gap between reviews of the investment.

[41] The investment was approved with the condition that the functional cost and risks be reviewed and reassessed by the end of the fiscal year, but there is no evidence the investment was reviewed again by the IT Program Management Board. The CIO Leadership Team was briefed on an exit strategy for this investment in May and July 2016, and the investment may have formally been elevated to that board for discussion, but neither board's meeting minutes documented such a decision.

[42] NASA's Office of the CIO reported that certain mission directorates had begun to have success in identifying previously unreported IT. However, as of the fiscal year 2018 budget, the CIO continued to have greater insight

2016 to explore how to identify such investments so that they could be reported to the CIO. The team initiated efforts to identify such investments in mission directorates and evaluated various mechanisms that NASA could employ to detect unreported IT. However, the agency has not yet finalized decisions about how to implement the team's recommendations, including those for fully identifying investments for all mission directorates or determining which mechanisms to employ to identify unreported IT. According to NASA officials, time frames for completing these activities have not yet been established.

In July 2017, NASA officials, including the Deputy CIO, acknowledged in governance board meeting minutes describing needed improvements, that the agency had not yet fully identified its IT footprint and needed to establish a comprehensive investment management process to address federal requirements, including those governing processes for selecting, reselecting, and overseeing IT investments. NASA officials explained that important progress had been made in improving oversight practices, but that efforts to implement more thorough capital investment reviews and identify IT investments across the agency had not yet been completed. The officials reported that they anticipated additional improvement to be made by the next annual budget cycle.

However, expanding NASA's oversight of IT will require continued coordination with the mission directorates to work through any needed changes to the longstanding differences in NASA's management of mission and business IT. The scope and complexity of such efforts are likely to be significant and may take time to plan and implement. Clearly defining how IT across the agency is to be identified and reported to the CIO would likely involve changes to policies and processes within and across NASA's IT, engineering, and mission program areas and would involve expertise and collaboration from those same groups. Until such practices are fully established, NASA will continue to operate with limitations in its oversight process and projects that fall short of performance expectations. In addition, the agency will face increased risk that its oversight will fail to (1) prevent duplicative investments, (2) identify opportunities to improve efficiency and effectiveness, and (3) ensure that investment progress and performance meet expectations.

NASA Has Not Yet Fully Defined Policies and Procedures for Managing Investments as a Portfolio

The IT investment management framework developed by GAO notes that, as investment management processes mature, agencies move from project specific processes to managing investments as a portfolio. The shift from investment management to IT portfolio management enables agencies to evaluate potential investments by how well they support the agency's missions, strategies, and goals. According to the framework, the

---

and influence over investments in agency IT services and space communications and navigation than in other mission IT investments.

investment board enhances the IT investment management process by developing a complete investment portfolio. As part of the process to develop a complete portfolio, an agency is to establish and implement policies and procedures for developing the portfolio criteria, creating the portfolio, and evaluating the portfolio.

NASA has not yet fully defined its policies and procedures for developing the portfolio criteria, creating the portfolio, and evaluating the portfolio. In its *Annual Capital Investment Review Implementation Plan*, dated October 2015, NASA began documenting policies for IT portfolio management and procedures for creating and evaluating the portfolio. For example, the procedures state that NASA is to update its IT portfolio annually in conjunction with the agency's planning and budgeting process. Additionally, in its *IT Capital Planning and Investment Control Process* guide, dated October 2006, NASA outlined procedures the agency can use to analyze the portfolio by establishing factors that should be taken into consideration, including the relative benefits, costs, and risks of the investment compared to all other proposals and the strength of the investment's linkage to NASA's strategic business plan.

However, these documents do not constitute a comprehensive IT portfolio management process in that they do not specifically define the procedures for creating and modifying the IT portfolio selection criteria; analyzing, selecting, and maintaining the investment portfolio; or reviewing, evaluating, and improving the performance of its portfolio. Further, the policies and procedures have not been updated to reflect current NASA practices. Specifically, the current policies and procedures have not been updated to reflect changes the agency made to its capital investment review process that are relevant to portfolio management.

According to NASA officials, the reason that the agency has not fully defined its policies and procedures is because they are intended to be part of a new IT portfolio management framework that also requires NASA to make changes to its investment management process. Specifically, the IT portfolio management plan that NASA drafted in January 2017 called for the agency to develop new IT investment criteria, discover currently unreported IT investments, develop an investment review process, and implement an IT investment dashboard and reporting tool, and a communications plan.

Although the IT Council has not yet approved the IT portfolio management plan, NASA has begun work to address elements of the draft plan, including building the requirements for an IT dashboard and reporting tool for implementation in 2018. In addition, according to Office of the CIO officials, the capital planning team is continuing to work with stakeholders to develop a comprehensive IT framework and investment review process. However, no firm dates have been established for the approval and implementation of the final plan or the framework. Until NASA fully defines its policies and procedures for developing the portfolio criteria, creating the portfolio, and evaluating the portfolio, the agency will lack assurance it is identifying and selecting the appropriate mix of IT projects that best meet its mission needs.

NASA Has Not Fully Established an Effective Approach for Managing Cybersecurity Risk

We have previously reported that securing federal government computerized information systems and electronic data is vital to the nation's security, prosperity, and well-being. Yet, the security over these systems is inconsistent and agencies have faced challenges in establishing cybersecurity approaches. Accordingly, we have recommended that federal agencies address control deficiencies and fully implement organization-wide information security programs.

NIST's cybersecurity framework is intended to support federal agencies as they develop, implement, and continuously improve their cybersecurity risk management programs.[43] In this regard, the framework identifies cybersecurity activities for achieving specific outcomes over the lifecycle of an organization's management of cybersecurity risk.[44] According to NIST, the first stage of the cybersecurity risk management lifecycle—which the framework refers to as "identify"—is focused on foundational activities for effective risk management that provide agencies with the organizational understanding to manage cybersecurity risk to systems, assets, data, and capabilities. NIST also provides specific guidance for implementing foundational activities and achieving desired outcomes that calls for, among other things, the following:[45]

- A risk executive in the form of an individual or group that provides agency-wide oversight of risk activities and facilitates collaboration among stakeholders and consistent application of the risk management strategy.
- A cybersecurity risk management strategy that articulates how an agency intends to assess, respond to, and monitor risk associated with the operation and use of the information systems it relies on to carry out the mission.
- An information security program plan that describes the security controls that are in place or planned for addressing an agency's risks and facilitating compliance with applicable federal laws, executive orders, directives, policies, or regulations.[46]

---

[43] NIST, Framework for Improving Critical Infrastructure Cybersecurity (Feb. 12, 2014).

[44] According NIST's cybersecurity framework, there are five stages in the cybersecurity risk management lifecycle: "identify," "protect," "detect," "respond," and "recover." They are intended to aid an organization in expressing its management of cybersecurity risk by organizing information, enabling risk management decisions, addressing threats, and improving by learning from previous activities.

[45] NIST, Security and Privacy Controls for Federal Information Systems and Organizations, Special Publication (SP) 800-53, Revision 4 (Gaithersburg, Md.: April 2013); Managing Information Security Risk: Organization, Mission, and Information System View, SP 800- 39 (Gaithersburg, Md.: March 2011).

[46] Security controls are the safeguards and countermeasures (management, operational, and technical) needed to protect the confidentiality, integrity, and availability of an information system and its information.

- Risk-based policies and procedures that act as the primary mechanisms through which current security requirements are communicated to help reduce the agency's risk of unauthorized access or disruption of services.
- However, NASA has not yet fully implemented these foundational activities of effective cybersecurity risk management.

Efforts to Establish Executive Oversight of Cybersecurity Are Underway

According to NIST guidance,[47] federal agencies should establish a risk executive in the form of an individual or group that provides organization-wide oversight of risk activities and facilitates collaboration among stakeholders and consistent application of the risk management strategy. This functional role helps to ensure that risk management is institutionalized into the day-to-day operations of organizations as a priority and integral part of carrying out missions.

NASA has developed a policy regarding the establishment of a risk executive function in accordance with NIST guidance, but it has not fully implemented the policy. Specifically, the agency's policy designates the Senior Agency Information Security Officer (SAISO) as the risk executive. According to the policy, the SAISO is charged with ensuring that cybersecurity is considered and managed consistently across the systems that support the agency and its partnerships—academic, commercial, international, and others that leverage NASA resources and extend scientific results.[48] The policy also calls for the SAISO to establish an office with the mission and resources for information security operations, security governance, and cyber-threat analysis.

In accordance with its policy, NASA has designated an Acting SAISO. Since April 2017, the Acting SAISO has led the IT Security Division within the Office of the CIO—an office that coordinates information security operations, security governance, security architecture and engineering, and cyber-threat analysis.

However, the agency has not yet established a risk executive office with assigned leadership positions and defined roles and responsibilities. According to NASA documentation, the agency had planned for the office to become operational by mid-December 2016. Agency officials, including the Acting Deputy Associate CIO for Information Security, explained that an IT security program office was not established in 2016 because the planned time frame for doing so was not realistic and failed to take into account other risk management efforts competing for available resources. For example, the officials stated that the agency was focused on a priority goal of deploying a centralized tool across its centers that would provide monitoring of implemented security controls to ensure they are functioning adequately.

---

[47] NIST 800-39.

[48] Systems that support NASA and its partnerships provide, for example, business operations, ground support, publicly accessible web applications, and spacecraft control and communications.

According to the NASA CIO, the agency planned to establish a comprehensive risk executive function by employing a cybersecurity risk manager in April 2018 and forming a program office—called the Enterprise Security Office—by September 2018. NASA's new cybersecurity risk manager began work on April 2, 2018. The agency's plan to have the new cybersecurity risk manager establish a comprehensive risk executive function should help ensure that current risk management efforts and decisions are appropriate and consistently carried out across the agency and its external partnerships.

NASA Has Not Yet Established an Agency-Wide Cybersecurity Risk Management Strategy

NIST guidance[49] states that federal agencies should establish and implement an organizational strategy for managing cybersecurity risk that guides and informs how the agency assesses, responds to, and monitors risk to the information systems being relied on to carry out its mission. The strategy should, among other things, make explicit an agency's risk tolerance,[50] accepted risk assessment methodologies, a process for consistently evaluating risk across the organization, risk response strategies, approaches for monitoring risk over time, and priorities for investing in risk management.

In 2015, NASA recognized the need to establish and implement an agency-wide strategy for managing its cybersecurity risks to address weaknesses it had identified with the decentralized approach it was using. Specifically, because the agency's centers had independently developed approaches for managing cybersecurity risk, there was little integration regarding risk management and practices across the agency. Further, NASA determined that the decentralized, center-level approach did not provide sufficient transparency regarding risks that could affect mission directorate programs.

To overcome the limitations of its decentralized approach, NASA planned to develop and begin implementing a comprehensive cybersecurity strategy by the end of September 2016 that was expected to include the key elements identified in NIST guidance. For example, it was expected to define the agency's risk tolerance, establish a methodology for identifying and assessing risks, and provide a clear understanding of NASA's risk posture.

However, the strategy was not completed as planned and is currently in development. According to officials in the Office of the CIO, including the Acting Deputy Associate CIO for Information Security, the strategy was not completed as planned due to the complexity and scope of the effort. For example, the officials stated that establishing an effective agency-wide strategy required insight into center-specific practices and significant input from stakeholders at all levels of NASA. In addition, these officials and the NASA CIO

---

[49] NIST, Framework for Improving Critical Infrastructure Cybersecurity (Feb. 12, 2014); SP 800-53, Revision 4; and NIST SP 800-39.

[50] Risk tolerance is the level of risk or degree of uncertainty that is acceptable to organizations. It affects the nature and extent of risk management oversight, the extent and rigor of risk assessments performed, and the context of organization strategies for responding to risk.

explained that the agency's efforts were redirected in order to respond to a new executive order from the President to develop an action plan for adopting NIST's cybersecurity framework in phases.[51]

According to NASA's CIO, the agency plans to move forward with drafting an agency-wide cybersecurity strategy that reflects the agency's approach to using NIST's framework; however, the agency has not yet established time frames for completing this effort. Until NASA establishes and implements a comprehensive strategy for managing its cybersecurity risks using NIST's framework, its ability to make operational decisions that adequately address security risks and prioritize IT security investments will be hindered.

NASA's Information Security Program Plan Does Not Fully Address Relevant Leading Practices and Is Not Finalized

NIST recommends that federal agencies develop and disseminate an information security program plan that describes the organization-wide security controls that are in place or planned for addressing the agency's risks and complying with applicable federal laws, executive orders, directives, policies, or regulations.[52] Specifically, the plan should provide a description of the agency's program management controls and common controls[53] in place or planned for meeting relevant federal, legal, or regulatory requirements; include the identification and assignment of roles, responsibilities, and coordination among organizational entities responsible for different aspects of information security; define the frequency for reviews of the security program plan; and receive approval from a senior official with responsibility and accountability for the risk being incurred.

NASA issued a draft information security program plan in November 2017 that addresses many of the components called for in NIST guidance. For example, the plan discusses

- program management controls that will be established, including the development of an inventory of its information systems, measures to determine information

[51] Exec. Order No. 13800, *Strengthening the Cybersecurity of Federal Networks and Critical Infrastructure*, 82 Fed. Reg. 22,391 (May 16, 2017). In responding to the executive order, NASA identified key issues that the agency needed to address in order to implement an integrated cybersecurity risk management approach. These issues include the lack of (1) a dedicated cybersecurity team to provide agency-level direction and recommendations, (2) cybersecurity workforce planning, and (3) effective processes for reporting cybersecurity risks.

[52] NIST, SP 800-53, Revision 4. According to NIST, security controls are the safeguards and countermeasures (management, operational, and technical) needed to protect the confidentiality, integrity, and availability of an information system and its information.

[53] Program management controls focus on organization-wide information security requirements that are independent of any particular information system and are essential for managing information security programs. Common controls, also referred to as inherited controls, provide a security capability for multiple information systems within an organization. When common controls are used to support a specific information system, they are referenced by that specific system as an inherited control.

security performance, and an information security workforce development and improvement program;

- common controls that are to be implemented agency-wide, including configuration management, contingency planning, and personnel security;[54]
- roles and responsibilities for promoting collaboration and providing consolidated unclassified security operations, and incident response and IT security awareness and training capabilities;[55] and
- responsibility for ensuring that the information security program plan is maintained, approved by the NASA CIO, and reviewed annually.

However, the plan is currently in draft and incomplete. For example, it does not yet describe the majority of the security functions and services that are to be carried out by the agency's IT Security Division to address the relevant federal statutory and regulatory requirements. Specifically, the plan does not identify the agency-wide privacy controls derived from standards promulgated pursuant to federal law and guidance that, according to the agency, are an integral part of its security program.[56]

According to NASA's Acting Deputy Associate CIO for Information Security, the information security program plan has not been finalized because of an upcoming revision to NIST's guidance for implementing security controls. Specifically, a fifth revision of NIST SP 800-53 is planned for release in December 2018. NASA's Acting Deputy Associate CIO for Information Security stated that the agency intends to finalize its draft plan after incorporating the updated NIST guidance.

In the absence of an established information security program plan, NASA's view of the security controls that protect its systems will remain decentralized, and it will lack assurance that it has established oversight over security controls for all of its systems. In addition, the agency will continue to operate its systems without defined and established information security requirements that are essential to agency-wide operations.

NASA's Security Policies and Procedures Are Not Always Current or Integrated

NIST Special Publication 800-53 recommends that agencies create policies and procedures to facilitate the appropriate application of security controls.[57] If properly

---

[54] Configuration management controls are intended to prevent unauthorized changes to information system resources (for example, software programs and hardware configurations) and to provide reasonable assurance that systems are configured and operating securely and as intended.

[55] Incident response controls are necessary for rapidly detecting incidents, minimizing loss and destruction, mitigating the weaknesses that were exploited, and restoring computing services.

[56] Privacy controls are the administrative, technical, and physical safeguards employed within organizations to protect and ensure the proper handling of personally identifiable information (for example, an individual's name, social security number, or biometric records).

[57] NIST, *Framework for Improving Critical Infrastructure Cybersecurity* (Feb. 12, 2014) and SP 800-53, Revision 4.

implemented, these policies and procedures may be able to effectively reduce the risk that could come from cybersecurity threats such as unauthorized access or disruption of services. Because risk-based policies and procedures are the primary mechanisms through which federal agencies communicate views and requirements for protecting their computing environments, it is important that they are established and kept current.

NASA has taken steps to document policies and procedures that address the security controls identified in NIST guidance for protecting information systems. For example, the agency established an overarching security policy that identified roles and responsibilities related to configuration management, contingency planning, and incident response. In addition, the agency issued procedures for implementing each of the NIST controls.

However, NASA does not have current and fully integrated policies and procedures. For example, the agency's overarching policy for implementing security controls expired in May 2017. In addition, approximately one-third of the documents that guide the implementation of these controls remained in effect past their expiration dates instead of being updated before they had expired per NASA policy requirements.

Further, in July 2017, NASA determined that cybersecurity roles and responsibilities were not always clear and sufficiently integrated across policies. For example, responsibilities were not consistently well-defined in the policies for governance, IT security, program and project management, and systems engineering. In addition, although *NASA's Policy Directive 2810.1E, NASA Information Security Policy* provided the SAISO with responsibility for the agency's cybersecurity risk, the policy assigned mission directorates control over risk decisions for their missions and programs and the centers were given the authority to implement any technical changes needed to address risk.

NASA's *Procedural Requirement 2810.1A, Security of Information Technology* states that the agency's SAISO is responsible for ensuring that information security policies and procedures are reviewed and appropriately updated. However, according to officials in the Office of the CIO, including the specialist for IT security, responsibilities for establishing, reviewing, and updating policies and procedures are being shared by two groups: the IT Security Division, led by the SAISO, and the Capital Planning and Governance Division. Specifically, the IT Security Division controls the content of IT-related policies and procedures but does not have control over the established NASA-wide process for reviewing the policies and procedures to determine if any changes are needed to the content. Instead, the Capital Planning and Governance Division is responsible for ensuring formal review and approval of any IT-related policies and procedures through the standard agency process and schedule.

Officials from the Office of the CIO, including the specialist for IT security, also stated that they intend to (1) establish a policy management framework that would provide the SAISO with more control over policies and procedures and include an annual document review, and (2) clarify and update cybersecurity roles and responsibilities in NASA policies. However, the agency has not yet developed a plan and specific time frame for completing these activities.

In addition, the Acting Deputy Associate CIO for Information Security stated that, having expired policies and procedures is not significant because they will remain in use until they are rescinded or superseded by updated versions. However, until NASA fully updates its policies and procedures to govern security over the agency's computing environments, it will have limited assurance that controls over information are appropriately applied to its systems.

CONCLUSION

NASA continues to pursue efforts to improve IT strategic planning, workforce planning, IT governance, and cybersecurity, but consistently lacks the documented processes needed to ensure that policies and leading practices are fully addressed. Specifically, the agency has taken steps to improve the content of its strategic plan and established an agency-wide goal for improving its workforce. In addition, after analyzing its IT management and governance structure, NASA took action to streamline its governance boards and standardize and strengthen its selection and oversight of investments, including initiating a portfolio management process. NASA has also moved toward new strategies and plans to bolster cybersecurity.

Nevertheless, while NASA has made progress, the agency has not yet fully addressed many of the leading IT management practices noted in this report or completed efforts to increase the CIO's authority over, and visibility into, agency-wide IT. Among other things, NASA has not fully documented a process for IT strategic planning or addressed all key elements of a comprehensive plan. In addition, it has not yet fully implemented a workforce planning process and has gaps in efforts to address leading practices. Regarding IT governance, its efforts to institute an effective governance structure and update policies and procedures for selecting IT investments are not yet complete. Moreover, NASA has not yet addressed weaknesses in its oversight practices or fully defined policies and procedures for developing an effective portfolio management process.

Similarly, although NASA continues cybersecurity improvement efforts, important elements of an effective cybersecurity approach have not been completed, including establishing a risk management strategy, an information security program plan, and updated policies and procedures. Until NASA leadership fully addresses these leading practices, its ability to overcome its longstanding weaknesses and ensure effective oversight and management of IT across the agency will remain limited. Moreover, NASA may be limited in its ability to strengthen its risk posture, including ensuring effective cybersecurity across partnerships with commercial entities, federal agencies, and other countries.

RECOMMENDATIONS FOR EXECUTIVE ACTION

We are making 10 recommendations to the National Aeronautics and Space Administration:

- Administrator should direct the Chief Information Officer to develop a fully documented IT strategic planning process, including methods by which the agency defines its IT needs and develops strategies, systems, and capabilities to meet those needs. (Recommendation 1)
- The Administrator should direct the Chief Information Officer to update the IT strategic plan for 2018 to 2021 and develop associated implementation plans to ensure it fully describes strategies the agency will use to achieve the desired results and descriptions of interdependencies within and across programs. (Recommendation 2)
- The Administrator should direct the Chief Information Officer to address, in conjunction with the Chief Human Capital Officer, gaps in IT workforce planning by fully implementing the eight key IT workforce planning activities noted in this report. (Recommendation 3)
- The Administrator should direct the Chief Information Officer to institute an effective IT governance structure by completing planned improvement efforts and finalizing charters to fully establish IT governance boards, clearly defining roles and responsibilities for selecting and overseeing IT investments, and ensuring that the governance boards operate as intended. (Recommendation 4)
- The Administrator should direct the Chief Information Officer to update policies and procedures for selecting investments to provide a structured process, including thresholds and criteria needed for, among other things, evaluating investment risks as part of governance board decision making, and outline a process for reselecting investments. (Recommendation 5)
- The Administrator should direct the Chief Information Officer to address weaknesses in oversight practices and ensure routine oversight of all investments by taking action to document criteria for escalating investments among governance boards and establish procedures for tracking corrective actions for underperforming investments. (Recommendation 6)
- The Administrator should ensure that the Chief Information Officer fully defines policies and procedures for developing the portfolio criteria, creating the portfolio, and evaluating the portfolio. (Recommendation 7)
- The Administrator should direct the Chief Information Officer to establish an agency-wide approach to managing cybersecurity risk that includes

- a cybersecurity strategy that, among other things, makes explicit the agency's risk tolerance, accepted risk assessment methodologies, a process for consistently evaluating risk across the organization, response strategies and approaches for monitoring risk over time, and priorities for risk management investments; (Recommendation 8)
- an information security program plan that fully reflects the agency's IT security functions and services and agency-wide privacy controls for protecting information; (Recommendation 9) and
- policies and procedures with well-defined roles and responsibilities that are integrated and reflect NASA's current security practices and operating environment. (Recommendation 10)

AGENCY COMMENTS AND OUR EVALUATION

We provided a draft of this product to NASA for comment. In its comments, which are reproduced in appendix II, NASA concurred with seven of the recommendations, partially concurred with two recommendations, and did not concur with one recommendation.

NASA partially concurred with our first and second recommendations. Specifically, consistent with the first recommendation, NASA agreed to fully document its strategic planning process, including the methods by which the agency defines IT needs and develops outcomes, strategies, major actions, and performance measures to meet those needs.

In addition, our second recommendation called for NASA to update the strategic plan and develop associated implementation plans. With regard to updating the plan, NASA stated that its strategic plan provides the context and parameters to support achievement of the agency's vision and mission through the strategic use of IT. The agency also stated that this plan describes the business outcomes, strategies, major actions, and performance measures to achieve the desired results.

With regard to the implementation plans related to our first and second recommendation, NASA agreed to develop the associated implementation plans for accomplishing the IT strategic plan, including descriptions of the interdependencies within and across programs. Nevertheless, in commenting on both recommendations, as well as the first recommendation, NASA stated that it does not believe that implementation plans, including specific IT capability and system changes, should be part of a strategic plan. The agency also maintained that the implementation plans, including descriptions of interdependencies within and across programs, are at a lower level than the IT strategic plan, since detailed IT implementation plans are more dynamic than the four-year NASA IT Strategic Plan.

However, our first and second recommendations do not call for NASA to incorporate implementation plans within the strategic plan. Rather, as discussed in the report, it is important that NASA document how it intends to accomplish the activities outlined in the strategic plan. Further, we continue to believe that NASA should address the weaknesses we identified in this report by updating the strategic plan to incorporate strategies on resources and time frames to achieve desired results and descriptions of interdependencies within and across projects so that they can be understood and managed. Thus, we stand by both recommendations (recommendations 1 and 2) that the agency take these actions.

NASA did not concur with our third recommendation to implement the IT workforce planning activities noted in our report. In this regard, the agency stated that its workforce improvement efforts were already underway. Specifically, NASA stated that IT workforce planning is part of the agencywide Mission Support Future Architecture Program. It added that, among other things, this program is intended to ensure that mission support resources, including the IT workforce, are optimally structured to support NASA's mission. In addition, NASA referenced our two additional ongoing audits of the agency's IT workforce, and noted that its activities related to IT workforce planning would be centered on any recommendations resulting from those audits.

In our view, neither of these circumstances should hinder NASA from addressing our recommendation in this report. As of March 2018, the agency's IT workforce plans were out-of-date and incomplete because activities the agency had been planning since 2015 had not been finalized in an approved plan or implemented. Further, NASA had not yet determined when the Office of the CIO would become an active part of the agencywide Mission Support Future Architecture program or developed plans for when that program's assessment of the IT workforce would be completed.

Thus, instead of limiting NASA's ability to address our recommendation, implementing the workforce planning activities discussed in this report could complement the agency's ongoing and future efforts. Specifically, NASA could use the IT workforce leading practices described in this report to strengthen any new workforce plans and assess the implementation of any planned improvements. Until NASA documents an IT workforce planning process and implements all of the key IT workforce planning activities, the agency may not be effectively positioned to anticipate and respond to changing staffing needs. Further, the agency is likely to face challenges in controlling human capital risks when developing, implementing, and operating IT systems.

NASA concurred with our four recommendations aimed at addressing deficiencies in its IT governance (recommendations 4 through 7). In this regard, the agency described planned actions intended to address each of these recommendations. For example, among other activities, the agency stated that it intended to publish charters for all IT governance boards; have the IT Council review governance board operations annually; document criteria for escalating investments among governance boards; and update policies and procedures for managing its investments as a portfolio.

Similarly, NASA concurred with our three recommendations related to establishing an agency-wide approach to managing cybersecurity risk (recommendations 8, 9, and 10). The agency described actions it had taken or planned to address each of these recommendations. In particular, with regard to establishing a cybersecurity risk management strategy (recommendation 8), NASA asserted that it had already taken actions that met the requirements of our recommendation. Specifically, NASA stated that it had established an approach to developing its cybersecurity risk management strategy by approving a charter for an agency-wide team to address cybersecurity risk management needs and hiring a Chief Cybersecurity Risk Officer to oversee agency-wide risk management initiatives.

While these actions constitute steps toward addressing the recommendation, we disagree that establishing a charter for a team and hiring a Chief Cybersecurity Risk Officer fully addresses the recommendation. As previously noted in this report, the agency does not have a cybersecurity risk management strategy that includes elements of NIST guidance. The strategy should, among other things, make explicit the agency's risk tolerance, accepted risk assessment methodologies, a process for consistently evaluating risk across the organization, risk response strategies, approaches for monitoring risk over time, and priorities for investing in risk management. Ensuring that the established agency-wide team and the Chief Cybersecurity Risk Officer develop a cybersecurity risk management strategy that aligns with the NIST guidance will be essential to fully address our recommendation.

APPENDIX I: OBJECTIVE, SCOPE, AND METHODOLOGY

The National Aeronautics and Space Administration Transition Authorization Act of 2017 included a provision for us to review the effectiveness of the agency's approach to overseeing and managing information technology (IT), including its ability to ensure that resources are aligned with agency missions, cost effective, and secure.[1] Our specific objective for this review was to address the extent to which the National Aeronautics and Space Administration (NASA) has established and implemented leading IT management practices in strategic planning, workforce planning, governance, and cybersecurity.

To address this objective, we compared NASA's IT management policies, procedures, and other documentation to criteria established by federal laws and leading practices.[2] This

[1] National Aeronautics and Space Administration Transition Authorization Act of 2017, Pub. L. No. 115-10, §811(b), 131 Stat. 18, 59 (March 21, 2017).

[2] ²Federal Cybersecurity Workforce Assessment Act of 2015, Pub. L. No. 114-113, Div. N, Title III, 129 Stat. 2242, 2975-77 (Dec. 18, 2015); Carl Levin and Howard P. "Buck" McKeon National Defense Authorization Act for Fiscal Year 2015, Div. A, Title VIII, Subtitle D—Federal Information Technology Acquisition Reform, Pub. L. No. 113-291, § 835 128 Stat. 3292, 3449 (Dec. 19, 2014), codified at 41 U.S.C. § 1704 note; E-Government Act of 2002, Pub. L. No. 107-347, § 209, 116 Stat.2899, 2923-32 (Dec. 17, 2002), codified at 44 U.S.C. § 3501 note;

documentation included the agency's strategic plans, workforce gap assessments, governance board meeting minutes and briefings, charters, policies and procedures, and other documentation of the Chief Information Officer's (CIO) authority. We also reviewed relevant reports by GAO and the NASA Office of Inspector General.[3]

With regard to IT strategic planning, we identified the strategic plans and related planning guidance issued by NASA and the Office of the CIO, including *NASA's Governance and Strategic Management Handbook*, dated November 26, 2014; *NASA's Information Resources Management Strategic Plan*, dated March 2014; and NASA's updated *Information Technology Strategic Plan* for fiscal years 2018 to 2021. We then reviewed the agency's overall strategic plan, and evaluated its previous and current IT strategic plans against key practices for IT strategic planning that we have previously identified.[4] These practices call for documenting the agency's IT strategic planning processes and developing an IT strategic plan that

- aligns with the agency's overall strategy;
- identifies the mission of the agency, results-oriented goals, and performance measures that permit the agency to determine whether implementation of the plan is succeeding;
- includes strategies the governing IT organization will use to achieve desired results; and
- provides descriptions of interdependencies within and across projects so that they can be understood and managed.

---

and Clinger-Cohen Act of 1996, Pub. L. No. 104- 106, Div. D and Div. E, § 5125(c)(3), 110 Stat. 642, 684-85 (Feb. 10, 1996), codified at 40 U.S.C. § 11315(c)(3). Office of Management and Budget, *Circular No. A-11: Preparation, Submission, and Execution of the Budget*, July 2017; OMB *Circular No. A-130: Managing Information as a Strategic Resource,* (Washington, D.C.: July 28, 2016); and OMB *Memorandum M-13-09 Fiscal Year 2013 PortfolioStat Guidance: Strengthening Federal IT Portfolio Management* (Washington, D.C.: Mar. 27, 2013); Office of Personnel Management, *The Human Capital Framework* (https://www.opm.gov/policy-dataoversight/human-capital-management/); and GAO, *Human Capital: Key Principles for Effective Strategic Workforce Planning*, GAO-04-39 (Washington, D.C.: Dec. 11, 2003), and *Information Technology Investment Management: A Framework for Assessing and Improving Process Maturity*, GAO-04-394G (Washington, D.C.: March 2004).

[3] GAO, *Information Security: NASA Needs to Remedy Vulnerabilities in Key Networks*, GAO-10-4 (Washington, D.C.: Oct. 15, 2009) and National Aeronautics and Space Administration Office of Inspector General, *NASA's Efforts to Improve the Agency's Information Technology Governance*, IG-18-002 Washington, D.C.: Oct. 19, 2017) and *NASA's Information Technology Governance*, IG-13-015 (Washington, D.C.: Jun. 5, 2013).

[4] Leading practices were identified related to strategic planning from the Office of Management and Budget, Circular No. A-11: Preparation, Submission, and Execution of the Budget, July 2017; OMB Circular No. A-130: Managing Information as a Strategic Resource, (Washington, D.C.: July 28, 2016); and OMB Memorandum M-13-09 Fiscal Year 2013 PortfolioStat Guidance: Strengthening Federal IT Portfolio Management (Washington, D.C.: Mar. 27, 2013). Further, prior GAO work related to IT strategic planning and management practices includes, for example, GAO, Social Security Administration: Improved Planning and Performance Measures Are Needed to Help Ensure Successful Technology Modernization, GAO-12-495 (Washington, D.C.: Apr. 26, 2012); and Library of Congress: Strong Leadership Needed to Address Serious Information Technology Management Weaknesses, GAO-15-315 (Washington, D.C.: Mar. 31, 2015).

To determine the extent to which NASA has established and implemented leading IT workforce planning practices, we conducted a comparative analysis of NASA's IT workforce planning policies and documents. Specifically, we compared agency documents, such as NASA policy directives, the desk guide, and documentation of efforts to establish IT workforce competencies and staffing requirements and conduct gap assessments, to GAO's IT workforce framework.[5] GAO's framework consists of four IT workforce planning steps and eight key activities. The eight key activities were identified in federal law, regulations, and guidance, including the Clinger-Cohen Act of 1996,[6] the legislation referred to as the Federal Information Technology Acquisition Reform Act,[7] Office of Management and Budget (OMB) guidance,[8] the Office of Personnel Management's Human Capital Framework,[9] and GAO reports.[10]

Based on our assessment of the documentation and discussions with agency officials, we assessed the extent to which the agency implemented, partially implemented, or did not implement the activities. We considered an activity to be fully implemented if NASA addressed all of the underlying practices for the activity; partially implemented if it addressed some but not all of the underlying practices for the activity; and not implemented if it did not address any of the underlying practices for the activity.

We assessed IT governance practices by comparing NASA documentation to critical processes identified by GAO in the IT investment management framework.[11] To align our work with the provision in Section 811(a) of the National Aeronautics and Space Administration Transition Authorization Act of 2017 calling for NASA to take actions regarding IT governance,[12] we selected critical processes from Stage 2 of the framework:

- instituting the investment board;
- selecting and reselecting investments that meet business needs; and

---

5 GAO, IT Workforce: Key Practices Help Ensure Strong Integrated Program Teams; Selected Departments Need to Assess Skill Gaps, GAO-17-8 (Washington, D.C.: Nov. 30, 2016).

6 Pub. L. No. 104-106, Div. D and Div. E, § 5125(c)(3) (Feb. 10, 1996), codified at 40 U.S.C. § 11315(c)(3).

7 Federal Information Technology Acquisition Reform provisions of the Carl Levin and Howard P. 'Buck' McKeon National Defense Authorization Act for Fiscal Year 2015, Pub. L. No. 113-291, div. A, title VIII, subtitle D, 128 Stat. 3292, 3438-3450 (Dec. 19, 2014).

8 OMB, Management and Oversight of Federal Information Technology, Memorandum M15-14 (Washington, D.C.: June 10, 2015); Chief Information Officer Authorities, Memorandum M-11-29 (Washington, D.C.: August 8, 2011); Guidance for Specialized Information Technology Acquisition Cadres (Washington, D.C.: July 13, 2011).

9 OPM, Human Capital Framework, (Washington, D.C.: Dec. 2016). The framework is accessible at https://www.opm.gov/policy-data-oversight/human-capital-management/.

10 GAO, IT Workforce: Key Practices Help Ensure Strong Integrated Program Teams; Selected Departments Need to Assess Skill Gaps, GAO-17-8 (Washington, D.C.: Nov. 30, 2016); Human Capital: Key Principles for Effective Strategic Workforce Planning, GAO-04-39 (Washington, D.C.: Dec. 11, 2003).

11 GAO, Information Technology Investment Management: A Framework for Assessing and Improving Process Maturity, GAO-04-394G (Washington, D.C.: March 2004).

12 National Aeronautics and Space Administration Transition Authorization Act of 2017, Pub. L. No. 115-10, §811(a), 131 Stat. 18, 58 (2017).

- providing investment oversight.

For each critical process, we compared key practices outlined in the framework to NASA documentation. The documentation we reviewed included NASA's IT governance policies and procedures, and charters and other guidance. We also reviewed governance board meeting minutes and briefings from each board's first meeting in 2016 through meetings held in August 2017.

In addition, we selected key practices for effective governance from Stage 3 of the IT investment management framework regarding establishing and implementing policies and procedures for developing the portfolio criteria, creating the portfolio, and evaluating the portfolio. We then compared documentation, including NASA's *IT Capital Planning and Investment Control Process* guide dated October 2006, and *Annual Capital Investment Review Implementation Plan* dated October 2015, and draft IT portfolio management plans, against these practices.

Using standards and guidance from the National Institute of Standards and Technology (NIST),[13] which identify foundational elements of effective cybersecurity risk management, we evaluated NASA's cybersecurity risk management approach by

- analyzing policies and plans for establishing a comprehensive risk executive function;
- evaluating documents and plans for establishing a cybersecurity risk management strategy;
- comparing a draft Information Security Program Plan to determine if it was consistent with NIST guidance; and
- analyzing policies and procedures to determine if they address relevant NIST security controls and are current.

In addition to assessing NASA headquarters, we reviewed IT management practices at two of the agency's nine centers (Marshall Space Flight Center in Huntsville, Alabama; and Johnson Space Center in Houston, Texas) and at one of NASA's four mission directorates (the Human Exploration and Operations Mission Directorate). The two centers and one mission directorate were selected because they had the largest fiscal year 2017 IT budgets, respectively, as reported on the federal IT dashboard.[14] We also visited the Goddard Space Flight Center in Greenbelt, Maryland, because of the center's proximity to GAO. The results of our work at the selected NASA centers and mission directorate are not generalizable to other NASA centers and mission directorates.

---

[13] NIST, Framework for Improving Critical Infrastructure Cybersecurity (Feb. 12, 2014).

[14] GAO, IT Dashboard: Agencies Are Managing Investment Risk, but Related Ratings Need to Be More Accurate and Available, GAO-14-64 (Washington, D.C.: Dec 12, 2013). The dashboard is OMB's public website that reports performance and supporting data for major IT investments.

To assess the reliability of these data, we compared them to budgetary data obtained directly from NASA's Office of the CIO. We found the data to be sufficiently reliable for the purpose of identifying the NASA centers and mission directorate with the largest IT budgets. We also interviewed cognizant officials with responsibilities for IT management at NASA headquarters and for the selected centers and mission directorate.

We conducted this performance audit from May 2017 to May 2018 in accordance with generally accepted government auditing standards. Those standards require that we plan and perform the audit to obtain sufficient, appropriate evidence to provide a reasonable basis for our findings and conclusions based on our audit objectives. We believe that the evidence obtained provides a reasonable basis for our findings and conclusions based on our audit objective.

APPENDIX II: COMMENTS FROM THE NATIONAL AERONAUTICS AND SPACE ADMINISTRATION

National Aeronautics and Space Administration
Headquarters
Washington, DC 20546-001

May 1, 2018

Reply to Attn of: Office of the Chief Information Officer

Ms. Carol C. Harris
Director
Information Technology Acquisition Management Issues
United States Government Accountability Office
Washington, DC 20548

Dear Ms. Harris:

The National Aeronautics and Space Administration (NASA) appreciates the opportunity to review and comment on the Government Accountability Office (GAO) draft report entitled, "NASA Information Technology: Urgent Action Needed to Address Significant Management and Cybersecurity Weaknesses" (GAO-18-337), dated March 21, 2018.

In 2014, NASA acknowledged the need to change its culture and management of information technology (IT) and established the Business Services Assessment (BSA) for IT. We continue to implement recommendations resulting from the BSA assessment, establishing a strong foundation to manage IT as a strategic resource. NASA's efforts have resulted in many successes, including consolidation of network operations and operational funding, a more effective IT investment review process integrated with the Agency budget process, and fully embracing the intent of Executive Order (EO) 13800, "Strengthening the Cybersecurity of Federal Networks and Critical Systems Infrastructure."

NASA recognizes the importance of cyber risk management and increasing insight into NASA's IT systems and assets to enable adequate identification, mitigation planning, and risk reduction of potential cybersecurity vulnerabilities. To support these efforts, NASA recently hired a Cyber Risk Manager to advance cybersecurity efforts and is establishing an Enterprise Cybersecurity Services Office. At this time, NASA has completed 90 percent of Phase I continuous diagnostics and mitigation (CDM) tool deployment, expecting completion of Phase I by December 2018. NASA has also started CDM Phase II. NASA's work to develop its response to EO 13800 provided an additional foundational piece for improving NASA's cybersecurity posture. In response to the EO, the NASA Acting Administrator charted the Cybersecurity Executive order Tiger Team (CEOTT), who designed and conducted NASA's first-ever Agency-wide cybersecurity capability gap assessment. As a result, the CEOTT identified key cybersecurity risk management priorities, redefined the way leadership conceptualizes cybersecurity, and

laid the groundwork for incorporating cyber risks into the enterprise risk management framework to provide stronger insights to NASA's senior leadership.

Significant progress has been made in transforming the Agency's IT governance model to become more effective and efficient. The IT Council (ITC), with executive members from each mission directorate, Center, and functional area was established. The ITC approved the recent NASA IT Strategic Plan, providing first ever cross-Agency stakeholder and customer ownership. Although effective until fiscal year 2021, the IT strategic plan is a living document that NASA will amend to meet changing needs. In addition, NASA continues our work in formalizing IT programs, managing and provisioning more IT services at the enterprise level, and identifying legacy IT and determining approaches to prioritize modernization efforts.

In the draft report, GAO makes ten recommendations to the NASA Administrator intended to address the deficiencies identified in IT strategic planning, workforce planning, governance, and cybersecurity.

Specifically, GAO recommends the following:

Recommendation 1: The Administrator should direct the Chief Information Officer to develop a fully documented IT strategic planning process, including methods by which the agency defines its IT needs and develops strategies, systems, and capabilities to meet those needs.

Management's Response: NASA partially concurs with this recommendation. The Agency will fully document its IT strategic planning process, including methods by which NASA defines its IT needs and develops outcomes, strategies, major actions, and performance measures to meet those needs. NASA does not believe that implementation plans, including specific IT capability and system changes, should be part of a strategic plan since these are at a lower level than an Agency IT strategic plan. Detailed IT implementation plans are more dynamic than the four-year NASA IT Strategic Plan. The Agency will document the process for implementing the NASA IT Strategic Plan, including the approach for planning for IT capabilities and systems.

Estimated Completion Date: October 31, 2018.

Recommendation 2: The Administrator should direct the Chief Information Officer to update the IT strategic plan for 2018 to 2021 and develop associated implementation plans to ensure it fully describes strategies the agency will use to achieve the desired results and descriptions of interdependencies within and across programs.

Management's Response: NASA partially concurs with this recommendation. The NASA IT Strategic Plan provides the context and parameters to support achievement of the Agency's vision and mission through the strategic use of IT. This plan describes the business outcomes, strategies, major actions, and performance measures to achieve the desired results. NASA maintains that the implementation plans, including descriptions of interdependencies within and across programs, are at a lower level than IT strategic plan since detailed IT implementation plans are more dynamic than the four-year NASA IT Strategic Plan. NASA will develop the associated implementation plans to accomplish the NASA IT Strategic Plan, including descriptions of interdependencies within and across programs.

Estimated Completion Date: January 18, 2019.

Recommendation 3: The Administrator should direct the Chief Information Officer address, in conjunction with the Chief Human Capital Officer, gaps in IT workforce planning by fully implementing the eight key IT workforce planning activities noted in this report.

Management's Response: NASA does not concur with this recommendation. NASA recognizes work is underway in the area of IT workforce planning. Currently, NASA is conducting a comprehensive, Agency-wide effort called the Mission Support Future Architecture Program (MAP). This effort is designed to ensure that NASA mission support resources, including workforce, are optimally structured to achieve the NASA mission. MAP will transform mission support services while maintaining mission focus, improving efficiency, ensuring local authority, and valuing the workforce. Simply put, MAP's goal is to optimize mission support services by moving toward a more interdependent model and freeing up resources to re-invest in facilities, IT, and other capabilities necessary for achieving NASA's ambitious portfolio of missions. In addition, GAO has two ongoing audits of IT workforce and recommendations from those audits would be in conjunction with this work. NASA recommends continued progress in the MAP effort and activities related to IT workforce planning remain centered in recommendations from the two GAO IT workforce audits.

Recommendation 4: The Administrator should direct the Chief Information Officer to institute an effective IT governance structure by completing planned improvement efforts and finalizing charters to fully establish IT governance boards, clearly defining roles and responsibilities for selecting and overseeing IT investments, and ensuring that the governance boards operate as intended.

Management's Response: NASA concurs with this recommendation. Charters will be published for all IT governance boards. The six IT program plans will define roles and responsibilities for selecting and overseeing IT investments. The Agency ITC will conduct an annual review of the governance board operations.

Estimated Completion Date: January 31, 2019

Recommendation 5: The Administrator should direct the Chief Information Officer to update policies and procedures for selecting investments to provide a structured process, including thresholds and criteria needed for, among other things, evaluating investment risks as part of governance board decision making, and outline a process for re-selecting investments.

Management's Response: NASA concurs with the recommendation. The Office of the Chief Information Officer guidance for Strategic Programming Guidance for fiscal year 2020 articulates how to document and report the Agency's IT investment information to enable the appropriate analysis of IT investments. The ITC will review and approve the Agency's IT portfolio of investments as part of this existing Program, Planning, Budgeting and Execution (PPBE) process in the summer of 2018, prior to the Agency's budget submission in September.

Estimated Completion Date: January 31, 2019.

Recommendation 6: The Administrator should direct the Chief Information Officer to address weaknesses in oversight practices and ensure routine oversight of all investments by taking action to document criteria for escalating investments among governance boards and establish procedures for tracking corrective actions for underperforming investments.

Management's Response: NASA concurs with the recommendation. NASA is documenting the criteria for escalating IT investments among governance boards and establishing procedures for tracking corrective actions for underperforming IT investments.

Estimated Completion Date: January 18, 2019.

Recommendation 7: The Administrator should ensure that the Chief Information Officer fully defines policies and procedures for developing the portfolio criteria, creating the portfolio, and evaluating the portfolio.

Management's Response: NASA concurs with the recommendation. NASA is updating policies and procedures for developing the portfolio criteria, creating the portfolio, and evaluating the portfolio.

Estimated Completion Date: February 28, 2019.

Recommendation 8: The Administrator should direct the Chief Information Officer to establish an agency-wide approach to managing cybersecurity risk that includes a cybersecurity strategy that, among other things, makes explicit the agency's risk tolerance, accepted risk assessment methodologies, a process for consistently evaluating risk across the organization, response strategies and approaches for monitoring risk over time, and priorities for risk management investments.

Management's Response: NASA concurs with the recommendation. NASA has established an approach to developing its cybersecurity risk management strategy which it believes meets the requirements for this recommendation. This approach includes (1) approving a charter for an Agency-wide Cybersecurity Integration Team to address key cybersecurity risk management needs; and (2) hiring a Chief Cybersecurity Risk Officer specifically to establish and oversee Agency-wide cybersecurity risk management initiatives. NASA completed these actions on April 2, 2018, as its approach to developing its cybersecurity risk management strategy and can provide documentation as needed.

Completion Date: April 2, 2018.

Recommendation 9: The Administrator should direct the Chief Information Officer to establish an agency-wide approach to managing cybersecurity risk that includes an information security program plan that fully reflects the agency's IT security functions and services and agency-wide privacy controls for protecting information.

Management's Response: NASA concurs with the recommendation. NASA is in the process of updating its information security program plan in accordance with the National Institute of Standards and Technology Special Publication 800-53 Revision 4, Security and Privacy Controls for Federal Information Systems and Organizations.

Estimated Completion Date: October 18, 2018.

Recommendation 10: The Administrator should direct the Chief Information Officer to establish an agency-wide approach to managing cybersecurity risk that includes policies and procedures with well-defined roles and responsibilities that are integrated and reflect NASA's current security practices and operating environment.

Management's Response: NASA concurs with the recommendation. NASA is reviewing and updating its policy management framework to facilitate consistent reviews and updates based on current security practices and the current operating environment.

Estimated Completion Date: March 29, 2019.

Thank you for the opportunity to comment on the subject draft report. If you have any questions or require additional information, please contact Ruth McWilliams on (202) 358-5125.

Sincerely,

Renee P. Wynn
Chief Information Officer

APPENDIX III: ACCESSIBLE DATA

Agency Comment Letter

Text of Appendix III: Comments from the Office of Federal Student Aid, Department of Education

Page 1

Dear Ms. Harris:

The National Aeronautics and Space Administration (NASA) appreciates the opportunity to review and comment on the Government Accountability Office (GAO) draft report entitled, "NASA Information Technology: Urgent Action Needed to Address Significant Management and Cybersecurity Weaknesses" (GAO-18-337), dated March 21, 2018.

In 2014, NASA acknowledged the need to change its culture and management of information technology (IT) and established the Business Services Assessment (BSA) for IT. We continue to implement recommendations resulting from the BSA assessment, establishing a strong foundation to manage IT as a strategic resource. NASA's efforts have resulted in many successes, including consolidation of network operations and operational funding, a more effective IT investment review process integrated with the Agency budget process, and fully embracing the intent of Executive Order (EO) 13800, "Strengthening the Cybersecurity of Federal Networks and Critical Systems Infrastructure."

NASA recognizes the importance of cyber risk management and increasing insight into NASA's IT systems and assets to enable adequate identification, mitigation planning, and risk reduction of potential cybersecurity vulnerabilities. To support these efforts, NASA recently hired a Cyber Risk Manager to advance cybersecurity efforts and is establishing an Enterprise Cybersecurity Services Office. At this time, NASA has completed 90 percent of Phase I continuous diagnostics and mitigation (CDM) tool deployment, expecting completion of Phase I by December 2018. NASA has also started CDM Phase II. NASA's work to develop its response to EO 13800 provided an additional foundational piece for improving NASA's cybersecurity posture.

In response to the EO, the NASA Acting Administrator charted the Cybersecurity Executive order Tiger Team (CEOTT), who designed and conducted NASA's first-ever Agency-wide cybersecurity capability gap assessment. As a result, the CEOTT identified key cybersecurity risk management priorities, redefined the way leadership conceptualizes cybersecurity, and laid the groundwork for incorporating cyber risks into the enterprise risk management framework to provide stronger insights to NASA's senior leadership.

Page 2

Significant progress has been made in transforming the Agency's IT governance model to become more effective and efficient. The IT Council (ITC), with executive members from each mission directorate, Center, and functional area was established. The ITC approved the recent NASA IT Strategic Plan, providing first ever cross-Agency stakeholder and customer ownership. Although effective until fiscal year 2021, the IT strategic plan is a living document that NASA will amend to meet changing needs. In addition, NASA continues our work in formalizing IT programs, managing and provisioning more IT services at the enterprise level, and identifying legacy IT and determining approaches to prioritize modernization efforts.

In the draft report, GAO makes ten recommendations to the NASA Administrator intended to address the deficiencies identified in IT strategic planning, workforce planning, governance, and cybersecurity.

Specifically, GAO recommends the following:

Recommendation 1: The Administrator should direct the Chief information Officer to develop a fully documented IT strategic planning process, including methods by which the agency defines its IT needs and develops strategies, systems, and capabilities to meet those needs.

Management's Response: NASA partially concurs with this recommendation. The Agency will fully document its IT strategic planning process, including methods by which NASA defines its IT needs and develops outcomes, strategies, major actions, and performance measures to meet those needs. NASA does not believe that implementation plans, including specific IT capability and system changes, should be part of a strategic plan since these are at a lower level than an Agency IT strategic plan. Detailed IT implementation plans are more dynamic than the four-year NASA IT Strategic Plan. The

Agency will document the process for implementing the NASA IT Strategic Plan, including the approach for planning for IT capabilities and systems.

Estimated Completion Date: October 31, 2018.

Recommendation 2: The Administrator should direct the Chief Information Officer to update the IT strategic plan for 2018 to 2021 and develop associated implementation plans to ensure it fully describes strategies the agency will use to achieve the desired results and descriptions of interdependencies within and across programs.

Management's Response: NASA partially concurs with this recommendation. The NASA IT Strategic Plan provides the context and parameters to support achievement of the Agency's vision and mission through the strategic use of IT. This plan describes the business outcomes, strategies, major actions, and performance measures to achieve the desired results. NASA maintains that the implementation plans, including descriptions of interdependencies within and across programs, are at a lower level than IT strategic plan since detailed IT implementation plans are more dynamic than the four-year NASA IT Strategic Plan. NASA will develop the associated implementation plans to accomplish the NASA IT Strategic Plan, including descriptions of interdependencies within and across programs.

Page 3

Estimated Completion Date: January 18, 2019.

Recommendation 3: The Administrator should direct the Chief Information Officer address, in conjunction with the Chief Human Capital Officer, gaps in IT workforce planning by fully implementing the eight key IT workforce planning activities noted in this report.

Management's Response: NASA does not concur with this recommendation. NASA recognizes work is underway in the area of IT workforce planning. Currently, NASA is conducting a comprehensive, Agency-wide effort called the Mission Support Future Architecture Program (MAP). This effort is designed to ensure that NASA mission support resources, including workforce, are optimally structured to achieve the NASA mission. MAP will transform mission support services while maintaining mission focus, improving efficiency, ensuring local authority, and valuing the workforce. Simply put, MAP's goal is to optimize mission support services by moving toward a more interdependent model and freeing up resources to re-invest in facilities, IT, and other capabilities necessary for achieving NASA's ambitious portfolio of missions. In addition, GAO has two ongoing audits of IT workforce and recommendations from those audits would be in conjunction with this work. NASA recommends continued progress in the MAP effort and activities related to IT workforce planning remain centered in recommendations from the two GAO IT workforce audits.

Recommendation 4: The Administrator should direct the Chief Information Officer to institute an effective IT governance structure by completing planned improvement efforts and finalizing charters to fully establish IT governance boards, clearly defining roles and responsibilities for selecting and overseeing IT investments, and ensuring that the governance boards operate as intended.

Management's Response: NASA concurs with this recommendation. Charters will be published for all IT governance boards. The six IT program plans will define roles and responsibilities for selecting and overseeing IT investments. The Agency ITC will conduct an annual review of the governance board operations.

Estimated Completion Date: January 31, 2019

Page 4

Recommendation 5: The Administrator should direct the Chief Information Officer to update policies and procedures for selecting investments to provide a structured process, including thresholds and criteria needed for, among other things, evaluating investment risks as part of governance board decision making, and outline a process for re-selecting investments.

Management's Response: NASA concurs with the recommendation. The Office of the Chief Information Officer guidance for Strategic Programming Guidance for fiscal year 2020 articulates how to document and report the Agency's IT investment information to enable the appropriate analysis of IT investments. The ITC will review and approve the Agency's IT portfolio of investments as part of this existing Program, Planning, Budgeting and Execution (PPBE) process in the summer of 2018, prior to the Agency's budget submission in September.

Estimated Completion Date: January 31, 2019.

Recommendation 6: The Administrator should direct the Chief Information Officer to address weaknesses in oversight practices and ensure routine oversight of all investments by taking action to document criteria for escalating investments among governance boards and establish procedures for tracking corrective actions for underperforming investments.

Management's Response: NASA concurs with the recommendation. NASA is documenting the criteria for escalating IT investments among governance boards and establishing procedures for tracking corrective actions for underperforming IT investments.

Estimated Completion Date: January 18, 2019.

Recommendation 7: The Administrator should ensure that the Chief Information Officer fully defines policies and procedures for developing the portfolio criteria, creating the portfolio, and evaluating the portfolio.

Management's Response: NASA concurs with the recommendation. NASA is updating policies and procedures for developing the portfolio criteria, creating the portfolio, and evaluating the portfolio.

Estimated Completion Date: February 28, 2019.

Recommendation 8: The Administrator should direct the Chief information Officer to establish an agency-wide approach to managing cybersecurity risk that includes a cybersecurity strategy that, among other things, makes explicit the agency's risk tolerance, accepted risk assessment methodologies, a process for consistently evaluating risk across the organization, response strategies and approaches for monitoring risk over time, and priorities for risk management investments.

Page 5

Management's Response: NASA concurs with the recommendation; NASA has established an approach to developing its cybersecurity risk management strategy which it believes meets the requirements for this recommendation. This approach includes (1) approving a charter for an Agency-wide Cybersecurity Integration Team to address key cybersecurity risk management needs; and (2) hiring a Chief Cybersecurity Risk Officer specifically to establish and oversee Agency-wide cybersecurity risk management initiatives. NASA completed these actions on April 2, 2018, as its approach to developing its cybersecurity risk management strategy and can provide documentation as needed.

Completion Date: April 2, 2018.

Recommendation 9: The Administrator should direct the Chief Information Officer to establish an agency-wide approach to managing cybersecurity risk that includes an information security program plan that fully reflects the agency's IT security functions and services and agency-wide privacy controls for protecting information.

Management's Response: NASA concurs with the recommendation. NASA is in the process of updating its information security program plan in accordance with the National Institute of Standards and Technology Special Publication 800- 53 Revision 4, Security and Privacy Controls for Federal Information Systems and Organizations.

Estimated Completion Date: October 18, 2018.

Recommendation 10: The Administrator should direct the Chief Information Officer to establish an agency-wide approach to managing cybersecurity risk that includes policies and procedures with well-defined roles and responsibilities that are integrated and reflect NASA's current security practices and operating environment.

Management's Response: NASA concurs with the recommendation. NASA is reviewing and updating its policy management framework to facilitate consistent reviews and updates based on current security practices and the current operating environment.

Estimated Completion Date: March 29, 2019.

Page 5

Thank you for the opportunity to comment on the subject draft report.

Sincerely,
Renee P. Wynn
Chief Information Officer

In: NASA: Major Projects and Select Issues
Editor: Oliver Cummings
ISBN: 978-1-53614-817-6

Chapter 5

NASA CONTRACTOR WHISTLEBLOWERS: STEPS TAKEN TO IMPLEMENT PROGRAM BUT IMPROVEMENTS TO TIMELINESS AND GUIDANCE NEEDED*

United States Government Accountability Office

WHY GAO DID THIS STUDY

NASA obligated over $18 billion in contracts and more than $1 billion in grants in fiscal year 2016, and it relies on a significant number of contractor and grantee employees to accomplish its work. These employees are legally protected from reprisal—such as demotion or firing—for whistleblowing.

GAO was asked to review NASA's whistleblower reprisal protections for contractor and grantee employees. This report addresses, among other things, the extent to which (1) NASA's Inspector General investigated contractor and grantee whistleblower reprisal complaints; (2) NASA's Administrator reviewed reprisal complaints in a timely manner; and (3) NASA communicated the applicable whistleblower reprisal protections to contractors. GAO reviewed NASA and its Inspector General's policies and guidance; reviewed a generalizable sample of 100 contracts from all NASA centers with contracts in fiscal year 2016; and interviewed relevant officials and contractors, grantees, and advocacy groups.

---

* This is an edited, reformatted and augmented version of United States Government Accountability Office; Report to the Ranking Member, Committee on Homeland Security and Governmental Affairs, U.S. Senate; Accessible Version, Publication No. GAO-18-262, dated March 2018.

WHAT GAO RECOMMENDS

GAO is making three recommendations to NASA, including evaluating the process for reviewing reprisal complaints to ensure it is meeting the required timeframe and clarifying guidance on when protections apply to contractor employees. NASA agreed with the recommendations and plans to develop a documented process to ensure it reviews reprisal complaints in a timely manner and clarify guidance as appropriate, among other things.

WHAT GAO FOUND

From 2008 through June 2017, National Aeronautics and Space Administration (NASA) contractor and grantee employees submitted 48 reprisal complaints such as alleged firing or demotion for reporting fraud, waste, or abuse within the government. NASA's Inspector General addressed all 48 complaints, completed investigations for 6 of those complaints, and forwarded investigation reports to the NASA Administrator, who is responsible for making a final determination of whether reprisal occurred. The Administrator determined that none of the complaints qualified for protection under the law.

Further, in 5 of the 6 cases forwarded by the OIG, the Administrator was required by statute to make a final determination of reprisal within 30 days. GAO found that the Administrator did not meet this required time frame for all 5 cases and had no documented response for one of them (see figure for all 5 cases). According to officials from NASA's Office of General Counsel, each case must be handled on a case by case basis to ensure due process and 30 days is insufficient time to issue an order of final determination of reprisal. However, in order to ensure that whistleblower reprisal complaints are handled within required timeframes, NASA would have to monitor and evaluate its processes for making final determinations of reprisal, but it has not yet taken this step. Consequently, NASA does not know what changes may be needed to ensure that it is meeting the statutory 30-day requirement.

NASA communicates whistleblower protections to contractors through inclusion of a required contract clause. For example, GAO found that almost all—98 percent—of contracts would be expected to include a whistleblower clause as required by statute. However, certain elements of NASA whistleblower protection guidance have contributed to a different understanding of reprisal protections among officials at headquarters, a NASA center, and the Inspector General. For example, a July 2014 procurement notice and contract clause language resulted in different interpretations about when the protections apply. Federal internal control standards require that an entity should communicate necessary quality information internally to meet the objectives of its mission. Without additional clarity in its guidance on when the protections apply, NASA centers and procurement officials will be at risk of inconsistent implementation of the law.

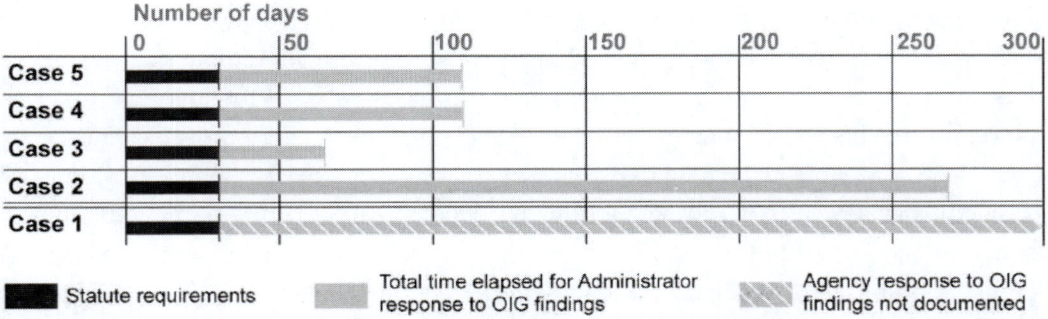

Source; GAO analysis of documents provided by National Aeronautics and Space Administration Office of Inspector General (OIG) and Administrator. | GAO-18-262.

Figure. Timeline for NASA Administrator's Response for 5 Investigated Reprisal Cases Subject to Statutory 30-Day Final Determination from 2008 through June 2017.

Abbreviations

FAR	Federal Acquisition Regulation
FPDS-NG	Federal Procurement Data System - Next Generation
JSC	Johnson Space Center
NASA	National Aeronautics and Space Administration
NSSC	NASA Shared Services Center
NFS	NASA FAR Supplement
NFS clause	NFS 1852.203-71
OIG	Office of Inspector General

441 G St. N.W.
Washington, DC 20548

March 8, 2018

The Honorable Claire McCaskill
Ranking Member
Committee on Homeland Security and Governmental Affairs
United States Senate
Dear Senator McCaskill:

When contractor and grantee employees report fraud, waste, abuse, and mismanagement, agencies should ensure they are prepared to fully investigate reprisal complaints—such as demotion or firing—for whistleblowing. Agencies also need to make a formal determination on whether reprisal occurred. Employees of contractors and grantees conduct a significant amount of work for the National Aeronautics and Space Administration (NASA). Given the sometimes enormous safety risks involved in space exploration, having an established process in place to investigate reprisal claims helps ensure NASA is able to act quickly whenever a whistleblower reports experiencing reprisal. In fiscal year 2016, NASA obligated over $18 billion in contracts to support acquisitions such as satellites and the construction of a solar plant; and more than $1 billion in grants for things such as science, technology, engineering, and math education initiatives and a student stipend for space-related research.

Although some protections for contractor employee whistleblowers have existed for years, section 2409 of Title 10, U.S. code (the statute), was amended in 2008, 2013, and 2014 to give contractor and grantee employee whistleblowers at NASA enhanced protection from reprisal. The statute establishes that an employee of a contractor, subcontractor, grantee, subgrantee, or personal services contractor may not be subject to reprisal for disclosing certain whistleblower complaints to certain authorities.[1]

You asked us to evaluate NASA's implementation of whistleblower reprisal protections. This chapter addresses the extent to which (1) NASA's Inspector General has investigated contractor and grantee whistleblower reprisal complaints, including meeting statutory timeliness requirements, and developed guidance for the investigations; (2) NASA's Administrator meets statutory timeliness requirements to review reprisal complaints; (3) NASA communicated externally and internally about the applicable contractor whistleblower reprisal protections; and (4) NASA communicated with grantees the applicable whistleblower reprisal protections.

To assess the extent to which NASA's Office of Inspector General (OIG) investigated reprisal complaints and developed guidance for the investigations, and the NASA Administrator met timeliness requirements, we obtained NASA guidance and relevant documentation. We conducted interviews with agency officials, including officials from the OIG and the Office of General Counsel on behalf of the Administrator about their processes and practices to conduct duties outlined in the statute. We reviewed data provided by the NASA OIG on the total number of whistleblower allegations of fraud, waste, abuse, misconduct, or mismanagement and reprisal complaints; the number of contractor and grantee employee whistleblower allegations and reprisal complaints; and the outcomes or decisions reached by the OIG of a reprisal complaint from fiscal years

[1] In this chapter, we use the term "contractors and grantee employees" to also include both subcontractors and subgrantees.

2008, the year the enhanced protections were put in place, through 2017.[2] We assessed the reliability of these data by asking the NASA OIG to describe the source(s) of information used and steps taken to identify the numbers provided, and limitations and caveats that would affect our use of these data—such as the data being self-reported by the OIG and Administrator. Based on these steps, we determined the data to be sufficient for our purposes of determining the extent to which the complaints were addressed.

To assess the extent to which NASA communicates whistleblower protections externally, to contractors and grantees, we identified new fiscal year 2016 contract awards over $300,000 as reported in the Federal Procurement Data System-Next Generation (FPDS-NG). We selected contracts that were not only over the simplified acquisition threshold (generally $150,000), but were over $300,000 to account for possible exceptions and to ensure that we were sampling contracts that would be required to include a whistleblower reprisal clause. To determine whether the required NASA Federal Acquisition Regulation (FAR) Supplement clause 1852.203-71 (NFS clause) or other potentially applicable clauses were included in NASA contracts, we selected for review a generalizable random sample of 100 contracts from a total of 270 contracts.[3] The sample is projectable to NASA fiscal year 2016 contracts; however, making a legal determination of compliance would require a case by case review, which we did not do for contracts not in our sample. The 100 contracts consist of a random selection of 10 contracts from each NASA center that awarded new contracts in fiscal year 2016; for those centers that did not have 10 contracts, we selected all contracts.[4] The remaining contracts were pulled from NASA Shared Services Center (NSSC), because that center does the most contracting. We estimated the percent of NASA contracts expected to include whistleblower clause(s) as the weighted average of results from the 10 contracting centers. Additionally, we selected a non-generalizable random sample of 10 grants awarded by NASA in fiscal year 2016 for review to determine whether NASA grants included a mechanism notifying grantee employees of their whistleblower rights and reprisal protections.[5] For both the sample of contracts and grants, we conducted data reliability checks by comparing the data to the documentation and concluded the data were sufficiently reliable for our purposes to determine if relevant whistleblower reprisal protections information was communicated externally.

To assess internal communication about whistleblower protections, we reviewed relevant documentation, including guidance, and conducted interviews with procurement

---

[2] While OIG officials track whistleblower reprisal complaints, they do not specifically track disclosures as a separate category. The numbers are based on a search of their database using key terms such as "disclosure" or "contractor employee" in the complaint field. Officials pointed out that complaints made through the OIG hotline may qualify for whistleblower protections, but the complainant may not identify him or herself as a whistleblower or even want to be considered a whistleblower.

[3] FAR clauses 52.203-15, 52.203-17, and 52.212-4.

[4] NASA is comprised of 12 centers and 7 facilities located throughout the United States.

[5] We selected 10 random grants from 2016 with positive dollar values to avoid grants captured in FPDS-NG as negative numbers due to deobligations.

officials, NASA Headquarters, NASA's Office of General Counsel, and Chief Counsels at Johnson Space Center (JSC), NSSC, and Marshall Space Flight Center. We also conducted interviews with contractors, grantees, and whistleblower advocacy groups. For additional information on our scope and methodology, see appendix I.

We conducted this performance audit from March 2017 to March 2018 in accordance with generally accepted government auditing standards. Those standards require that we plan and perform the audit to obtain sufficient, appropriate evidence to provide a reasonable basis for our findings and conclusions based on our audit objectives. We believe that the evidence obtained provides a reasonable basis for our findings and conclusions based on our audit objectives.

BACKGROUND

For contractor and grantee employees at NASA, whistleblower protections have changed over time. For example, by statute, in 2007, NASA contractor employees were protected against reprisal if they disclosed information relating to a substantial violation of law related to a contract. However, in 2008, amendments to the whistleblower statute provided protections only to those contractor employees at NASA who reported "a substantial and specific danger to public health or safety." In 2013, the statute was amended again to include disclosures of gross mismanagement of a NASA contract or grant, a gross waste of Administration funds, and abuse of authority relating to a NASA contract or grant, or a violation of law, rule, or regulation related to a NASA contract or grant. In 2014, the statute was further amended with the only significant change to protect grantee and subgrantee employees. See Table 1 for detailed description of the 2008, 2013, and 2014 amendments of the statute.

Disclosure: An allegation to certain bodies and individuals made by an employee who believes he or she has witnessed certain wrongdoing, such as gross mismanagement or gross waste.

Source: GAO analysis based on 10 U.S.C. § 2409(a) | GAO-18-262.

Reprisal Complaint: Following a disclosure, a complaint that an employee has experienced reprisal as a result of the disclosure, such as demotion or discharge.

Source: GAO analysis based on 10 U.S.C. § 2409(b) | GAO-18-262.

Table 1. Selected Summary of Amendments to the Whistleblower Reprisal Protections for NASA Contractor and Grantee Employees

10 U.S.C. § 2409	After fiscal year 2008 National Defense Authorization Act (NDAA) amendments[a]	After fiscal year 2013 NDAA amendments[b]	After fiscal year 2014 NDAA amendments[c]	Summary of selected changes
Employees covered	Contractor employees, same as previous.	Contractor and subcontractor employees.	Contractor, subcontractor, grantee, and subgrantee employees.	Expanded protections to subcontractor, grantee, and subgrantee employees.
Protected whistleblower disclosures	A substantial and specific danger to public health or safety.	Same as 2008, plus gross mismanagement of NASA contract or grant, a gross waste of Administration funds, an abuse of authority relating to an Administration contract or grant, or a violation of law, rule, or regulation related to an Administration contract (including the competition for or negotiation of a contract) or grant.	Same as previous	Extended protected disclosures to include gross mismanagement, waste, or abuse of funds,among other things for NASA contractor and grantee employees.[d]
Length of investigation	180 days/no limit Office of Inspector General (OIG) and complainant may agree to additional time to investigate.	180 days/360 maximum OIG may request up to an additional 180 days, with complainant approval.	Same as previous	Limited OIG to an extension up to 180 daysin addition to the initial 180 day investigative time frame.
Agency head review	30 days	Same as previous	Same as previous	No change
Employee notification	No requirement for contractors to communicate whistleblower protections to their employees.	NASA must ensure that its contractors and subcontractors inform their employees in writing of the rights and remedies provided by the statute in the predominant language of the workplace.[e]	Same as previous Although grantees protections were added, no requirement to ensure grantees inform employees of rights and remedies provided by statute.	NASA now includes a NASA Federal Acquisition Regulation Supplement clause in contracts requiring that its contractors and subcontractors inform their employees of whistleblower protections.

Source: GAO analysis of amendments to 10 U.S.C. § 2409 | GAO-18-262.

[a]Pub. L. No. 110-181, § 846 (enacted January 28, 2008). 10 U.S.C. § 2409 also applies to the Department of Defense and Coast Guard. For purposes of this table, we selected information specific to NASA.

[b]Pub. L. No. 112-239, § 827 (enacted January 2, 2013; amendments effective 180 days after enactment).

[c]Pub. L. No. 113-291, § 856(a) (enacted December 19, 2014).

[d]Prior to the 2008 amendments, NASA contractor employees were protected for disclosing information relating to a substantial violation of law related to a contract (including the competition for or negotiation of a contract). 10 U.S.C. § 2409 (2007).

[e]The statute does not include a requirement to inform grantee and subgrantee employees of their whistleblower reprisal protections.

Requirements under the Current Statute

Under the current statute, the NASA Office of Inspector General and Administrator have different responsibilities. Since the 2014 amendments, contractor, subcontractor, grantee, and subgrantee employees are protected from reprisal if they disclose to certain persons or bodies information they reasonably believe is evidence of gross mismanagement of a federal contract or grant, a gross waste of federal funds, an abuse of authority relating to a federal contract or grant, a substantial and specific danger to public health or safety, or a violation of law, rule, or regulation related to a federal contract or grant. Additionally, contractor employees may make whistleblower disclosures to several entities, including a management official at the contractor. Figure 1 depicts the disclosure process and the complaint process.

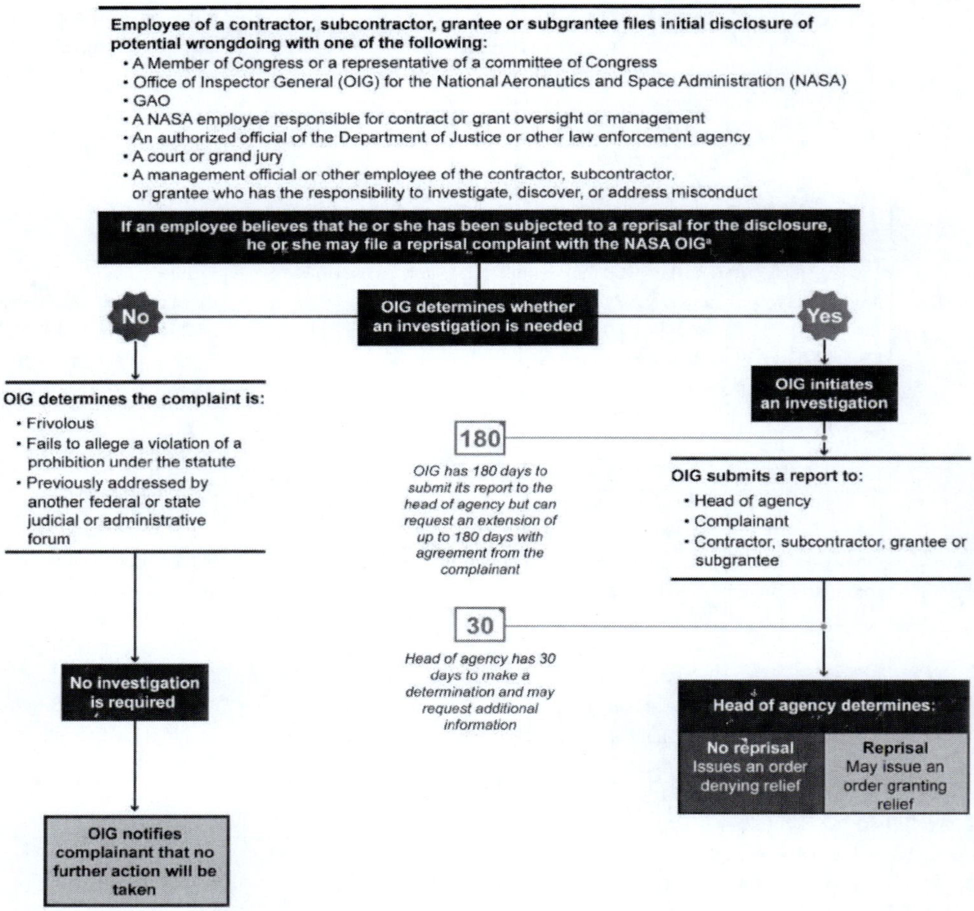

Source: GAO analysis of 10 U.S.C. § 2409. | GAO-18-262.

[a]An employee has three years from when the alleged reprisal took place to file a complaint.

Figure 1. Disclosure and Reprisal Complaint Investigation and Review Processes for NASA Contractor and Grantee Employees.

NASA OIG Role

Upon receiving a reprisal complaint, the OIG must evaluate whether a reprisal complaint is covered under the statute. In addition to the steps described in Figure 1 for investigating complaints, there are instances when the OIG does not investigate. The OIG might not investigate for a variety of reasons, such as in cases where the complaint is already under investigation by another authority such as another OIG, or otherwise does not allege a violation of the law, such as if whistleblower disclosure does not constitute gross fraud, waste, abuse or mismanagement. If the OIG determines the case is not covered under the statute, it may then notify the complainant that no further action will be taken on the reprisal complaint.

Administrator Role

Upon receipt of the NASA OIG investigation report, the NASA Administrator (the head of agency) has 30 days to determine whether the contractor made a prohibited reprisal and issue an order denying or granting relief. According to NASA officials, during the 30-day period after the agency head receives the OIG report, the agency practice has been to ask the OIG for any additional investigative work and also afford the complainant and the contractor an opportunity to submit a written response to the OIG report. Any person adversely affected or aggrieved by the administrator's order may, within 60 days of issuance, obtain a limited review by the U.S. circuit court of appeals.

Agency Procurement Official Role

Under the NFS regulations, NASA contracting officers are also responsible for inserting an NFS whistleblower clause into applicable contracts that requires contractors communicate to their employees their rights under the statute.[6] The NFS whistleblower clause lays out the responsibility of contractors to communicate to their employees their rights under the statute, in writing and in their predominant native language. All contracts over the simplified acquisition threshold awarded on or after July 29, 2014, require a whistleblower clause. The statute also requires NASA to make best efforts to include a clause providing for the applicability of the 2013 amendments in contracts awarded before July 1, 2013—the effective date of the 2013 amendments—that have undergone major contract modifications. The terms "best efforts" and "major modifications" are not defined in the statute.[7] Unlike provisions affecting contractors, the statute does not require NASA

[6] NFS 1852.203-71.

[7] Several different whistleblower protections exist in the FAR and agency supplements, with each applicable depending on considerations such as the agency in question, acquisition procedures, and source of funding. For some contracts, the FAR clause 52.203-17 is applicable. For all commercial item acquisitions, contracting officers must insert the FAR clause 52.212-4 that requires the contractor to comply with 10 U.S.C. § 2409. Contracts that obligate American Recovery and Reinvestment Act of 2009 funds require whistleblower protections unique to the Act found in FAR clause 52.203-15.

to ensure that grantees or subgrantees notify employees in writing of their rights under the statute.

NASA OIG Investigated Reprisal Complaints within Required Time Frames and Recently Updated Incomplete Guidance

From 2008 to June 2017, NASA OIG addressed whistleblower reprisal complaints within required time frames, according to OIG officials. At the time we initiated this review, the OIG's guidance for handling reprisal complaints had been updated to reflect most statutory changes; however, it did not include guidance regarding subgrantees. During the course of our review, the OIG updated the investigation guidance in October 2017 to include subgrantee employees.

NASA OIG Handled Reprisal Complaint Investigations within Required Time Frames

NASA OIG completed 6 reprisal investigations within required time frames.[8] The OIG received 277 whistleblower disclosures leading to 48 reprisal complaints from 2008 through June 2017, and handled those complaints within required time frames, according OIG officials.[9] For the 6 of those reprisal complaints that were investigated, the OIG used extensions. OIG officials said that extensions may be necessary for a number of reasons, including that the complaint may be highly technical in nature, requiring the OIG to find subject matter expertise to better understand the nature of the whistleblower complaint and whether it constitutes gross fraud, waste, abuse, or mismanagement.

When the OIG receives a reprisal complaint, complainants are asked to fill out a whistleblower complaint form and an investigation is initiated. See Figure 2 below for the process by which the OIG conducts its investigations.

In addition, there were 5 complaints currently under investigation and 37 complaints during this time frame that the NASA OIG did not investigate because the OIG deemed

---

[8] For those reprisal complaints covered under the 2013 amendments, the investigations must be completed within the time frame of 180 days, and an allowable, mutually agreed upon, one-time extension of 180 days for a total of up to 360 days.

[9] Reprisal complaints included instances of firing and demotion for reporting fraud, waste, or abuse.

them to be frivolous, determined they were not covered under the statute, or the complaint was handled in another forum, such as the court system or by another OIG.[10] Complaints were deemed frivolous for several reasons, including if the complainant did not want to disclose his or her identity and proceed with the claim, or the whistleblower disclosure happened after the reprisal. OIG officials told us that when cases are disposed of without an investigation, the OIG notifies the complainant of the decision in writing. Figure 3 shows the disposition of the 48 reprisal complaints received from 2008 through June 2017.

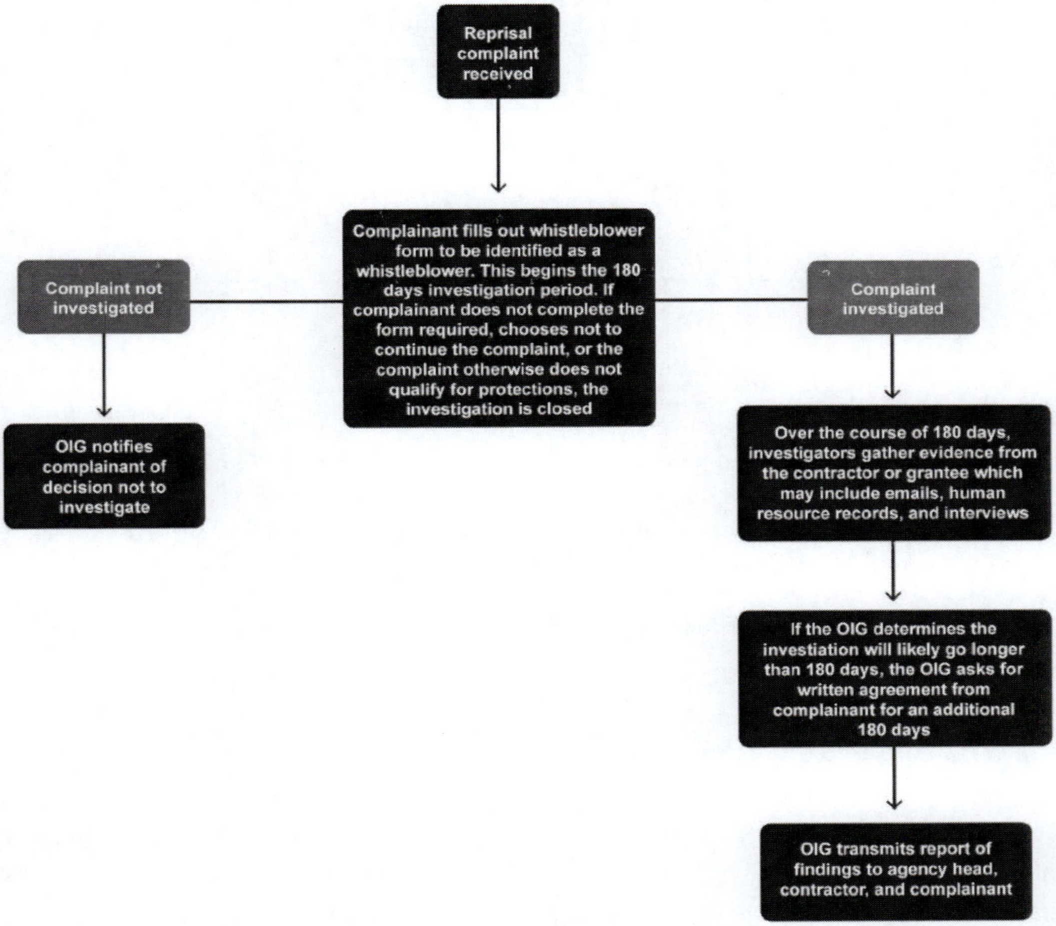

Source: GAO analysis of information and documentation received from National Aeronautics and Space Administration Office of Inspector General (OIG). | GAO-18-262.

Figure 2. NASA Office of Inspector General (OIG) Process for Investigating Reprisal Complaints.

[10] OIG officials said that all 37 complaints not investigated were closed before the 180 day time frame requirement for investigations.

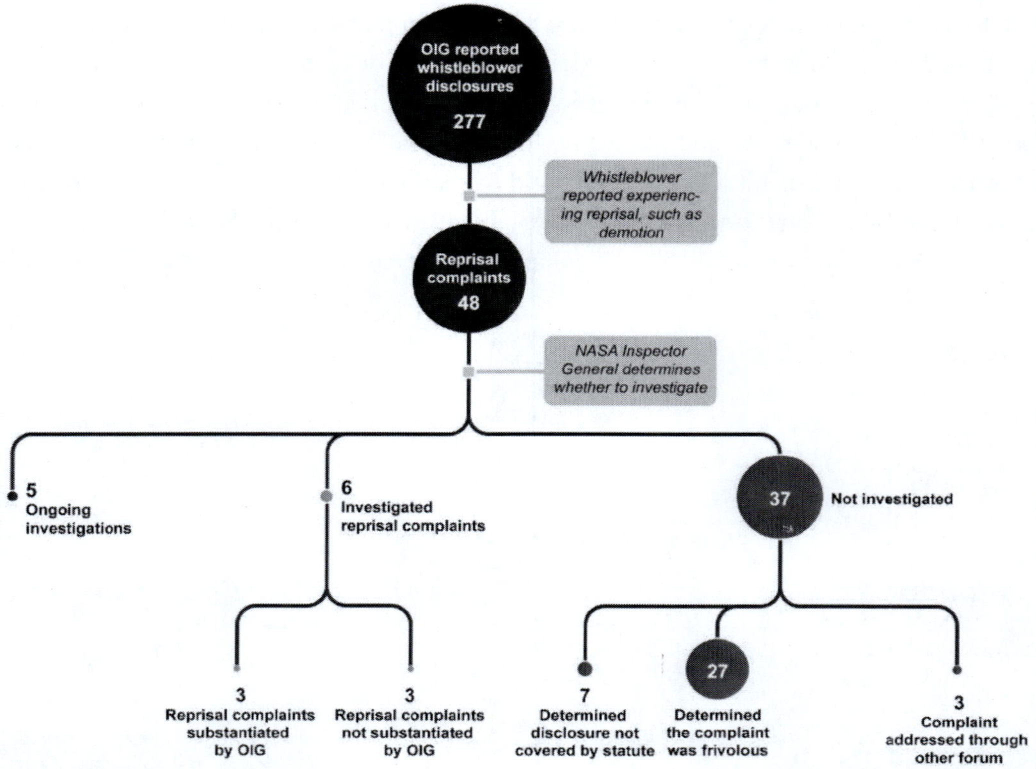

Source: GAO analysis of National Aeronautics and Space Administration's (NASA) Office of Inspector General (OIG) data. | GAO-18-262.

Note: Of the 48 reprisal complaints received by the OIG, one was made by the employee of a grantee; the rest were made by employees of contractors.

Figure 3. Disposition of Reprisal Complaints Received by NASA's Office of Inspector General (OIG) from 2008 through June 2017.

The OIG Updated Investigation Guidance

The OIG has developed guidance for conducting investigations, which includes a chapter on contractor and grantee whistleblower reprisal complaints. Although most changes to the statute (such as to whom reprisal may be reported) had been incorporated into the investigation guidance, the initial guidance provided to us by the OIG did not include a 2014 statutory requirement to extend protections to subgrantees. During the course of our review, in October 2017, the OIG updated its guidance for investigating reprisal complaints to include subgrantee employees. Because subgrantees are now protected by statute, including them in the investigation policy will help ensure they are consistently extended protections through OIG investigations.

In addition to its guidance, OIG officials said they have developed training specific to whistleblower investigations for new investigators, conducted internal training for investigators, and external training for contractor employees. Additionally, the OIG Investigators' Central Field Office conducts periodic training for investigators that includes any updates to whistleblower protections.[11] With regard to external training, the OIG officials said that investigators at some of the NASA centers—with the largest contract activity—have conducted on-site training for some contractors. This training is conducted as part of general fraud awareness training.

NASA Administrator Did Not Meet the Required Time Frames for Reprisal Complaint Review

The Administrator failed to meet the required review time frame and issue an order of final determination of reprisal for 5 completed investigations received from the OIG from 2008 through June 2017.[12] In all 5 cases, the Administrator took longer than the 30 days to issue an order. In one of those 5 complaints, an official from the Office of General Counsel was unable to provide us with the issued order and said he did not believe one was completed, and could not provide an explanation as to why an order was not completed. For the 5 reprisal complaints, Figure 4 shows the number of days from when the Administrator received an OIG report of findings to the time when an order of final determination was documented.

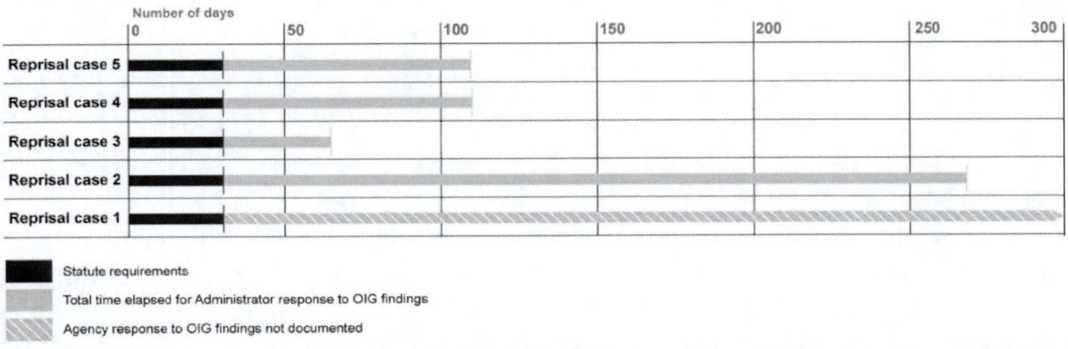

Source: GAO analysis of documents provided by National Aeronautics and Space Administration Office of Inspector General (OIG) and Administrator. | GAO-18-262.

Figure 4. Timeline for NASA Administrator's Response for 5 Investigated Reprisal Cases Subject to Statutory 30-Day Final Determination from 2008 through June 2017.

[11] The OIG Investigators' Central Field Office includes JSC, Marshall Space Flight Center, Kennedy Space Center, and Stennis Space Center.

[12] During this time period, the Administrator received 6 reports; however, only 5 were subject to the requirement that the Administrator make final determination of reprisal within 30 days. The sixth report of findings was based on a complaint made before the 2008 amendment to the statute was in effect; therefore, the 30-day requirement was not applicable.

In addition to the 5 complaints mentioned above, there was another OIG investigation of a reprisal complaint that did not require response from the administrator within 30 days, but was finalized within our review time frame, for a total of 6 completed OIG investigations of reprisal complaints. For 3 of the 6 complaints, the OIG found that reprisal had occurred and reported those findings to the administrator for final determination of reprisal. The Administrator determined that none of these 3 complaints qualified for protection under the law. For 2 of these complaints, the Administrator found that they did not qualify for protections because they fell under the 2008 version of the statute and failed to allege a violation specific to public health and safety.[13] In 2017, a court affirmed the Administrator's position.[14] For the third complaint, the Administrator determined reprisal could not be substantiated due to the complainant not meeting the standards of evidence under the statute.[15]

Further, we found that NASA does not have a standard process in place for the Administrator to review cases that qualify for protections under the statute and issue an order of final determination. According to an official from the Office of General Counsel, the agency has no standard process to ensure the contractors are afforded due process, among other things. The official from Office of General Counsel said the Administrator may need to conduct an additional investigation in some cases.[16] He said that each case is different and would have to be handled on a case by case basis. In addition, the official said the Administrator may need to conduct hearings, independent of the OIG. Moreover, the official from Office of General Counsel highlighted concern that the Administrator's office does not have the resources to conduct additional investigative work, which he said is a key contributor for the office's inability to meet the 30-day timeline to issue an order of final determination. Despite acknowledging these challenges, the Administrator does not have a formal process or criteria to monitor and evaluate the way the office handles issuing an order of final determination of reprisal to ensure that it meets the statutory time requirements.

Because the Administrator took longer than 30 days to respond to all reprisal complaints, including one where the Office of General Counsel failed to provide evidence

---

[13] After the 2008 amendments, 10 U.S.C. § 2409 provided protections to those contractor employees at NASA who disclosed "a substantial and specific danger to public health or safety." Whistleblower complaints made about general fraud, waste, abuse, and mismanagement alone, but not related to a substantial and specific danger to public health or safety, were not protected disclosures under the statue at that time.

[14] *Javery v. Bolden*, No. 16-60221, 2017 WL 2703790 (5th Cir. June 21, 2017).

[15] The Administrator determined the OIG report of findings and subsequent responses did not contain sufficient grounds to clearly support a finding that the complainant had been reprised against.

[16] NASA General Counsel explained that in its view there is not a one-size-fits-all answer to what process is due, because it is necessarily a fact-driven question. NASA General Counsel said its concerns stem from the Eighth Circuit's decision in *Business Communications, Inc. v. U.S. Dept. of Ed.*, 739 F.3d 374 (8th Cir. 2013). In that case, the Department of Education determined that a contractor had retaliated against a whistleblower in violation of section 1553 of the American Recovery and Reinvestment Act of 2009, which provides substantively similar whistleblower protections to the NASA Whistleblower statute. The contractor appealed. The court held that before the department could order relief for the whistleblower, it had to provide the contractor due process, including a hearing and the opportunity for the contractor organization to confront and cross examine adverse witnesses.

that the Administrator responded at all, there may be the unintended consequence of disincentivizing future whistleblowers from making disclosures who fear their complaint will not be handled timely. Internal controls require that management should establish and operate monitoring activities to monitor the internal control system, evaluate the results, and take appropriate action to address issues on a timely basis.[17] Without monitoring, evaluating, and taking appropriate corrective action based on the way the Administrator or his or her designee makes a final determination of reprisal, there is no assurance that whistleblower reprisal complaints will be handled within required time frames in the future.

NASA Almost Always Communicates Whistleblower Protections to Contractors, but Internal Guidance Is Unclear

In almost all of the contracts we reviewed, NASA had met its obligation to ensure its contractors are communicating whistleblower protections to their employees through a whistleblower contract clause. We also found that NASA has put in place guidance to its contracting workforce on the protections, and guidance on when to include the whistleblower clause in contracts. However, we found that some NASA officials have interpreted this guidance differently. Further, NASA's guidance does not reflect an agency-wide policy on when to include the whistleblower clause when modifying a contract.

Whistleblower Clause Included in Almost All Contracts, but a Few Were Missing the Required Clause

In most cases, NASA included a clause regarding whistleblower reprisal protections in applicable contracts to ensure contractors communicate rights to its employees. But we found that the clause was not included in all relevant contracts in our review. Based on our review of a generalizable sample of contracts, we estimate that 98 percent of contracts would be expected to include a whistleblower clause at the time new contracts were awarded in applicable contracts in 2016, and 2 percent would not.[18] Within our sample, 4 contracts did not have a whistleblower reprisal clause.[19] After we shared our contract file

---

[17] GAO, *Standards for Internal Control in the Federal Government*, GAO 14 704G (Washington, D.C.: September 2014).

[18] To assess legal compliance we would have to make a determination on case by case basis, which we did not do.

[19] Our review of a generalizable sample of 100 contracts found that 96 contracts in the sample included whistleblower language while 4 did not. The weighted estimate of the percentage of NASA contracts that would be expected to include whistleblower language is 97.7 percent, with a 95 percent confidence interval of 92.5 to 99.6 percent. The weighted estimate of the percentage of NASA contracts that would not be expected to include whistleblower language is 2.3 percent, with a 95 percent confidence interval of 0.4 to 7.5 percent. See Appendix I for more

review findings with NASA officials, they modified 3 of the 4 contracts to include the missing required whistleblower clause. For the remaining contract, the contractor performance was complete, the contract had been closed, and no further action will be taken.

According to NASA procurement officials, human error, combined with its former contract writing system, could explain why the contracts in our sample did not have the required clause. They explained that the former contracting writing system relied on templates and did not automatically include the NFS clause into all contracts. Under this system, contracting officers used templates that included a list of all potential or applicable NFS and FAR clauses, which are incorporated through a manual process. Officials said that if a clause were included in the templates, it is unlikely that it would be removed because doing so would require supervisory approval.

NASA procurement officials told us that the agency launched a new contract writing system in June 2017. They said that under the new contract writing system, contracting officers use a logic system that prepopulates each contract with required clauses. They said that the new automated system will likely lead to fewer human errors because inserting the clause will not be a manual process. Because the new system is still being implemented, we were unable to evaluate whether the risk of human error has been reduced or eliminated to ensure applicable contract awards have the clause.

Under the previous and current systems, NASA contracts are to undergo various levels of review prior to award—including attorney review—at the centers or headquarters based on risk level and dollar thresholds. For example, contracts awarded by JSC valued at over $50 million are to be reviewed by headquarters. NASA procurement officials stated that they conduct procurement management reviews, and centers conduct annual self-assessments; however, at one center, officials pointed out that these reviews have not previously included whether a whistleblower clause is included in new contracts or major modifications. They said this is because reviews typically focus on known issues or program risk, and inclusion of the whistleblower clause has not been previously identified as an issue or risk. Contractors we spoke with were generally aware of their responsibilities to communicate reprisal protections to their employees because their contracts with NASA included the required NFS whistleblower clause. In response to our review, NASA procurement officials said they plan to include a review of the inclusion of NFS whistleblower clause in future compliance reviews as an area of emphasis and will instruct centers to include whether the clause is included in applicable contracts as part of the centers' self-assessment process.

information on sampling error. Of the 96 contracts with a whistleblower clause, 84 had the NFS clause and 12 had an alternative whistleblower clause.

NASA's Guidance Contributes to Different Understanding of Reprisal Protections

Three elements of NASA's whistleblower reprisal protection guidance—its procurement notice, NFS clause, and definition of major modification— contribute to potential confusion or inconsistent application of whistleblower reprisal protections.

First, in July 2014, NASA notified its contracting officials of the changes to the NFS required by the 2013 amendments to section 2409 Title 10 of the U.S. code through a procurement notice 04-80, but this notice has been interpreted differently by officials in NASA Headquarters, a NASA center, and the OIG. Procurement notices are drafted by NASA Headquarters, reviewed and approved by NASA general counsel and NASA's Office of Procurement. The NASA centers, acting through their procurement offices and, as needed their legal counsel, review and implement the notices.

Table 2. Description of Two Whistleblower Reprisal Protection Clauses

Federal Acquisition Regulation (FAR) Clause 52.203-17	NASA FAR Supplement Clause 1852.203-71	Difference
Contractor Employee Whistleblower Rights and Requirement to Inform Employees of Whistleblower Rights (Apr. 2014) (a) This contract and employees working on this contract will be subject to the whistleblower rights and remedies in the pilot program on Contractor employee whistleblower protections established at 41 U.S.C. 4712 by section 828 of the National Defense Authorization Act for Fiscal Year 2013 (Pub. L. 112- 239) and FAR 3.908. (b) The Contractor shall inform its employees in writing, in the predominant language of the workforce, of employee whistleblower rights and protections under 41 U.S.C. 4712, as described in section 3.908 of the Federal Acquisition Regulation. (c) The Contractor shall insert the substance of this clause, including this paragraph (c), in all subcontracts over the simplified acquisition threshold.	Requirement To Inform Employees Of Whistleblower Rights (Aug. 2014) (a) The Contractor shall inform its employees in writing, in the predominant native language of the workforce, of contractor employee whistleblower rights and protections under 10 U.S.C. 2409, as described in subpart 1803.9 of the NASA FAR Supplement. (b) The Contractor shall include the substance of this clause, including this paragraph (b), in all subcontracts.	The NASA clause does not explicitly state that contractor employees will be subject to whistleblower remedies.

Source: GAO analysis of Federal Acquisition Regulation and National Aeronautics and Space Administration (NASA) FAR Supplement whistleblower protection clauses. | GAO-18-262.

After the procurement notice was issued, some NASA officials interpreted it differently. For example, in one instance, a NASA center Chief Counsel's office attorney advised a center procurement official that reprisal protections found in the 2013 amendments extend to contractors' employees working on contracts awarded before the effective date of the amendments. This is true, he said, regardless of whether the contract contains any clause explicitly making the 2013 amendments applicable. However, both the

OIG and the Administrator's counsel have expressed a different understanding of the statute conveyed in the notice, stating that a clause making the 2013 amendments applicable must be in a contract in order for the complainant to be protected under the statute. Later, the attorney from the center Chief Counsel's office revised his understanding of the statute and concluded the procurement notice was not accurate as written.

Second, NASA personnel have different understandings about whether the NFS clause is sufficient for contractor employees to be covered by the statute. The NFS clause instructs contractors to inform their employees in writing of contractor whistleblower employee rights; but, unlike the FAR clause that is used to implement similar legislation for other agencies, the NFS clause does not state that employees working on the contract are subject to the whistleblower rights and remedies. The attorney from a NASA center said that the NFS clause is enough to ensure contractor employees are given rights under the statute. However, OIG officials have said that without including that element of the clause, employees working under NASA contracts awarded prior to the effective date of the 2013 amendments may not be covered by those amendments. See Table 2 for description of the clauses and their differences.

Third, the lack of agency-wide guidance for when to include the clause in major modifications leads to different implementation of the requirement. The 2013 amendments require that NASA makes best efforts to include a whistleblower clause in contracts undergoing a major modification. NASA's July 2014 procurement notice also encourages contracting officers to include the NFS whistleblower clause when issuing major modifications to contracts awarded before July 29, 2014. However, it does not specify what a major modification is under this statute, and the statute itself does not define "major." According to NASA procurement officials at headquarters and at two NASA centers, it is at the discretion of the NASA Centers' offices of procurement and contracting officers to decide if a clause is inserted into modifications, and whether they are considered major. Procurement officials and the contracting officers we spoke with told us that there is no definition of major modifications in the law, regulation, or NASA Headquarters or Center policies or guidance. NASA procurement officials said this is because it could be different for each contract and the contracting officer makes the determination based upon the facts related to the situation. Nevertheless, without communicating the factors to consider when determining whether a modification is major and whether that contract should or should not include the clause, NASA and the Centers' procurement officials are at risk of potential inconsistent incorporation of the clause among applicable contracts.

One attorney in NASA's General Counsel's Office said there may be costs associated with asking a contractor to insert new clauses—such as the whistleblower clause—into an existing contract during a major modification because it would require a bilateral negotiation between the contractor and the agency. However, one contractor we spoke with said that there would be no cost to adding the clause and that doing so would not be an issue because the whistleblower clause is consistent with the internal whistleblower policies and practices of the institution. Further, he said that the institution he represents would be hesitant to argue against inclusion of the NFS clause in its contracts with NASA.

Internal control standards require that an entity should internally communicate necessary quality information in order to meet requirements of the mission. These 3 areas of potential confusion related to NASA's current guidance could result in different application of the law, unless they are clarified.

NASA Has Not Established a Mechanism to Communicate Whistleblower Protections to Grantees

Although whistleblower protections are now extended to grantee employees by statute, NASA does not have a mechanism in place to communicate the protections to grantees or subgrantee employees. Unlike the requirement for NASA to ensure contractors communicate whistleblower reprisal rights to their employees in writing and in the employees' predominant language, the statute does not prescribe a similar requirement for NASA to ensure that grantees communicate whistleblower reprisal rights to their employees. During the grant application process, NASA requires grantees to attest that they will not require grantee employees to sign confidentiality agreements that would prohibit them from reporting fraud, waste, and abuse.

NASA officials said that grant awards do not include a mechanism, such as a term or condition, to encourage NASA grantees to notify their employees of their whistleblower reprisal protections. In the 10 NASA grants from fiscal year 2016 that we reviewed, there was no requirement included for grantees to communicate these protections to employees. However, we found that all 10 grants included a statement that each award was subject to all applicable laws and regulations of the United States in effect on the date of award, including the Uniform Guidance. For federal grants in general, the Uniform Guidance provides a government-wide framework for grants management.[20] Within this guidance, there is a reference to the whistleblower protections in the statute; however, it does not explicitly describe the statute's requirements.[21]

The grant advocacy group and representatives of three NASA grantees we spoke with were aware that some protections exist and noted that many grantees have their own whistleblower policies, but were not aware of the specific protections provided by the statute, which indicates that opportunities exist for improving communications between NASA and its grantees about these protections. Further, representatives of the grant advocacy group noted that the whistleblower protections for grantee employees were not

[20] Office of Management and Budget produces the *Uniform Administrative Requirements, Cost Principles, and Audit Requirements for Federal Awards* also referred to as "Uniform Guidance."

[21] 2 CFR § 200.300(b).

specifically mentioned at recent annual meetings where grantees and federal officials discuss issues that affect grantees. Internal controls require that management externally communicate the necessary quality information to achieve the entity's objectives. Without additional communication about the protections provided by the statute, grantees may not fully understand or appreciate the significance of the rights afforded to their employees, and grantee employees may not be aware of their whistleblower reprisal protections, which could hinder their willingness to report instances of fraud, waste, and abuse.

CONCLUSION

Because contractor and grantee employee whistleblowers risk reprisal after disclosing potential fraud, waste, abuse and mismanagement, ensuring they are protected from retaliation or adverse consequences is critical. Without monitoring and evaluating the timeliness of reviewing and responding to reprisal complaints, the Administrator may not be prepared to determine reprisal on future cases within the statutorily required 30 days.

Additionally, although NASA has developed guidance related to contractor protections, this guidance has led to inconsistent interpretation of the law and could potentially lead to inconsistent application of how contractor protections for employees are conveyed. More clear guidance would help contracting officers determine when to incorporate the NFS clause into major modifications to ensure consistency throughout the agency.

Finally, because unlike contracts, there is no similar clause for grants, NASA is in the position to help ensure grantees know their employees' rights against reprisal if they observe and disclose fraud, waste, abuse or mismanagement. However, NASA has not effectively communicated to grantees information about these provisions and as a result grantees and their employees may not be fully aware of these protections. Consequently, if they witness fraud, waste, abuse or mismanagement, they may not be willing to disclose those for fear of reprisal.

RECOMMENDATIONS FOR EXECUTIVE ACTION

We are making three recommendations to NASA:

The Administrator should monitor, evaluate, and make appropriate corrective actions, such as a documented process, to ensure it reviews reprisal complaints in a timely manner. (Recommendation 1)

The Administrator should review NASA's guidance or develop other guidance, including defining major modification, to clarify when whistleblower protections are conveyed. (Recommendation 2)

The Administrator should communicate whistleblower protections to grantees and subgrantees and their employees. (Recommendation 3)

AGENCY COMMENTS AND OUR EVALUATION

We provided a draft of this chapter to NASA for review and comment. NASA provided written comments on a draft of this report. In its written comments, reprinted in appendix II, NASA concurred with all three recommendations. In its response to our recommendations NASA agreed to develop and document a process to ensure it reviews reprisal complaints in a timely manner to ensure all parties' due process rights are protected, review existing procurement policy and clarify guidance as appropriate, and update NASA grant guidance to communicate whistleblower protections to grantees, subgrantees and their employees. As agreed with your office, unless you publicly announce the contents of this report earlier, we plan no further distribution until 30 days from the report date. At that time, we will send copies to the appropriate congressional committees and members.

APPENDIX I: OBJECTIVES, SCOPE, AND METHODOLOGY

To assess the extent to which National Aeronautics and Space Administration's (NASA) Office of Inspector General (OIG) has investigated contractor and grantee whistleblower reprisal complaints and developed guidance for the investigations, we reviewed data provided by the NASA OIG on the total number of whistleblower allegations of fraud, waste, abuse, misconduct, or mismanagement and reprisal claims. We also reviewed the number of contractor and grantee employee whistleblower allegations and reprisal complaints provided by the OIG and the outcomes or decisions reached by the OIG of a reprisal complaint from fiscal years 2008 through 2017.[22] We assessed the reliability of these data by asking the NASA OIG to describe the source(s) of information used and steps taken to identify the numbers provided, and limitations and caveats that would affect GAO's use of the data—such as the data being self-reported by the OIG and Office of

---

[22] While OIG officials track whistleblower reprisal complaints, they do not specifically track disclosures as a separate category. The numbers are based on a search of their database using key terms such as "disclosure" or "contractor employee" in the complaint field. Officials pointed out that complaints made through the OIG's hotline may quality for whistleblower protections, but the complainant may not identify him or herself as a whistleblower or even want to be considered a whistleblower.

General Counsel. Based on these steps, we determined the data to be sufficient for our purposes of determining how the complaints were addressed. Additionally, we reviewed relevant documentation to assess the extent to which the NASA OIG was conducting investigations and communicating findings to the NASA Administrator within required time frames. To determine the extent NASA OIG developed guidance, we interviewed or obtained written answers from OIG officials about their processes and practices for investigating whistleblower reprisal complaints. We reviewed the guidance and training and other materials that NASA OIG uses to implement whistleblower protection investigations. We also visited Johnson Space Center (JSC)—selected because it had the highest number of reprisal cases from 2008 through 2017—to discuss policies and procedures specific to that center with OIG investigators and the OIG program manager for whistleblower protections. Because they are also a part of the Investigators' Central Field Office, we also spoke with investigators at Marshall Space Flight Center and Kennedy Space Center.

To assess the extent to which NASA's Administrator meets the statutory timeliness requirements to review reprisal complaints, we reviewed the timeliness of the Administrator's final determination to ensure that NASA was meeting statutory requirements. Specifically, we reviewed relevant documentation to assess the extent to which the Administrator was making final determination of reprisal in 30 days—the required review period specified by statute. We reviewed the Administrator's documentation on the final disposition of reprisal investigations and compared the date of the Administrator's final decision to the date of receipt of the reprisal investigations from the NASA OIG. We also conducted interviews with the Office of General Counsel, who spoke on behalf of the Administrator.

To assess the extent to which NASA communicated the applicable whistleblower reprisal protections externally with contractors, we reviewed a generalizable sample of NASA contracts to determine the extent a required whistleblower clause was included. We used the Federal Procurement Data System-Next Generation (FPDS-NG) to generate a sample of contract actions over $300,000 that were awarded by NASA in fiscal year 2016. We selected contracts that were not only over the simplified acquisition threshold (generally $150,000), but were over $300,000 to account for possible exceptions and to ensure that we were sampling contracts that would be required to include a whistleblower reprisal clause. From the 270 contracts identified, for purposes of examining the inclusion of NASA Federal Acquisition Regulation Supplement (NFS) clause 1852.203-71 (or other potentially applicable clauses) in NASA contracts, a legal requirement, we selected a generalizable random sample of 100 contracts. The sample is projectable to NASA fiscal year 2016 contracts; however, we did not make a case by case legal determination for contracts not in our sample. We randomly selected 10 contracts from each center that awarded new contracts in 2016, and for those centers that did not have 10 contracts, we selected all contracts.[23] The remaining contracts were then pulled from the NASA Shared Services Center (NSSC) because that center does a majority of NASA's contracting. We asked for contracts awarded in fiscal year 2016 to ensure we were sampling contracts that are required to have the clause and would be reasonably accessible by NASA. We excluded interagency contracts and task or delivery orders awarded using blanket purchase agreements to ensure we sampled base contracts awarded by NASA, not other agencies.

We estimated the percentage of NASA contracts expected to include whistleblower clause(s) as the weighted average of results from the 10 contracting centers. Because we followed a probability procedure based on random selections, our sample is only one of a large number of samples that could have been drawn. Because each sample could have provided different estimates, we express the uncertainty associated with any particular estimate as a 95 percent confidence interval. This is the interval that, with repeated sampling, would be expected to contain the actual population value for 95 percent of the samples we could have drawn. As a result, we are 95 percent confident that this confidence interval contains the true percentage of contracts expected to include whistleblower clause(s); however, to assess legal compliance we would have to make a case by case determination, which we did not do. We conducted data reliability checks on the FPDS-NG dataset by comparing it to contract documentation obtained from contract files and determined it was sufficiently reliable for our purposes. Additionally, we conducted interviews with NASA procurement officials and contracting officers at multiple locations including NASA Headquarters, NSSC and JSC to discuss any additional measures NASA takes to communicate whistleblower protections to its contractors and their contractor employees. To further assess internal communication, we reviewed relevant documentation, including guidance, and conducted interviews with procurement officials, NASA's Office of General Counsel, and Chief Counsels at JSC, NSSC, and Marshall Space Flight Center.

To assess the extent to which NASA communicated the applicable whistleblower reprisal protections with grantees, we reviewed a non-generalizable sample of grants awarded by NASA in fiscal year 2016 to determine whether NASA grants included a mechanism notifying grantees of their employees' whistleblower rights and reprisal protections. We used FPDS-NG to identify a non-generalizable random sample of 10 grants awarded by NASA in fiscal year 2016 for review to determine whether any of the selected grants included a mechanism to communicate whistleblower reprisal protections to grantee employees.[24] We conducted data reliability checks on the FPDS-NG data by comparing it to grant documentation obtained from grant awards and determined it was sufficiently reliable for our purposes. Additionally, we conducted interviews with NASA grant making

---

[23] NASA is comprised of 12 centers and 7 facilities located throughout the United States.

[24] We selected 10 random grants from grants awarded in 2016, with positive dollar values to avoid grants captured in FPDS-NG as negative numbers due to deobligations.

officials to discuss any additional measures NASA takes to communicate whistleblower reprisal protections to its grantees and their grantee employees.

Finally, in order to learn about contractor and grantee experiences during NASA's implementation of enhanced whistleblower protections, we conducted interviews with or received written answers to questions from a selected group of NASA contractors and grantees. Using FPDS-NG data, we selected institutions with the highest and lowest contracts (including small business contracts) and grants by obligations in 2016.[25] Using these selection criteria, we selected three contractors and three grantees to meet with based on the amount of funds obligated in 2016. We ultimately interviewed or obtained written answers from all selected contractors and grantees. Additionally, we spoke with two advocacy groups, one about grants and one about contracts.

We conducted this performance audit from March 2017 to March 2018 in accordance with generally accepted government auditing standards. Those standards require that we plan and perform the audit to obtain sufficient, appropriate evidence to provide a reasonable basis for our findings and conclusions based on our audit objectives. We believe that the evidence obtained provides a reasonable basis for our findings and conclusions based on our audit objectives.

APPENDIX II: COMMENTS FROM THE NATIONAL AERONAUTICS AND SPACE ADMINISTRATION

National Aeronautics and Space Administration
Office of the Administrator
Washington, DC 20546-0001

NASA

February 13, 2018

Ms. Shelby S. Oakley
Director
Contracting and National Security Acquisitions
United States Government Accountability Office
Washington, DC 20548

[25] "Low" grant obligations starts at $150,000. We selected a figure that is consistent with the simplified acquisition threshold. This number was also selected because it reflects an organization that still has some significant spending, and would therefore benefit more from being aware of these protections.

Dear Ms. Oakley:

The National Aeronautics and Space Administration (NASA) appreciates the opportunity to review and comment on the Government Accountability Office (GAO) draft report entitled, "NASA Contractor Whistleblowers: Steps Taken to Implement Program but Improvements to Timeliness and Guidance Needed" (GAO-18-262), dated January 18, 2018.

In the draft report, GAO makes three recommendations to the NASA Administrator intended to improve NASA's implementation of its contractor and grantee employee whistleblower program. Specifically, GAO recommends the following:

Recommendation 1: The Administrator should monitor, evaluate, and make appropriate corrective actions, such as a documented process, to ensure it reviews reprisal complaints in a timely manner.

Management's Response: Concur. NASA will develop a documented process to ensure it reviews reprisal complaints in a timely manner that ensure all parties due process rights are protected.

Estimated Completion Date: March 31, 2019

Recommendation 2: The Administrator should review NASA's guidance or develop other guidance, including defining major modification, to clarify when whistleblower protections are conveyed.

Management's Response: Concur. NASA will review existing procurement policy relative to when whistleblower protections are conveyed and provide clarifying guidance as appropriate.

Estimated Completion Date: October 31, 2018

Recommendation 3: The Administrator should communicate whistleblower protections to grantees and sub-grantees and their employees.

Management's Response: Concur. NASA will update the NASA Grant and Cooperative Agreement Manual to communicate whistleblower protections to grantees and sub-grantees and their employees.

Estimated Completion Date: October 31, 2018

Once again, thank you for the opportunity to comment on the subject draft report. If you have any questions or require additional information, please contact Dave Barrett on (202) 358-2074.

Sincerely,

Robert M. Lightfoot, Jr.
Administrator (Acting)

APPENDIX III: ACCESSIBLE DATA

Data Tables

Data Table for Highlights Figure, Figure: Timeline for NASA Administrator's Response for 5 Investigated Reprisal Cases Subject to Statutory 30-Day Final Determination from 2008 through June 2017

Statute requirements	Total time elapsed for Administrator response to OIG findings		Agency response to OIG findings not documented
Reprisal case 5	30	80	NA
Reprisal case 4	30	80	NA
Reprisal case 3	30	35	NA
Reprisal case 2	30	239	NA
Reprisal case 1	30	NA	243

Data Table for Figure 1: Disclosure and Reprisal Complaint Investigation and Review Processes for NASA Contractor and Grantee Employees

Employee of a contractor, subcontractor, grantee or subgrantee files initial disclosure of potential wrongdoing with one of the following:

- A Member of Congress or a representative of a committee of Congress
- Office of Inspector General (OIG) for the National Aeronautics and Space Administration (NASA)
- GAO
- A NASA employee responsible for contract or grant oversight or management
- An authorized official of the Department of Justice or other law enforcement agency
- A court or grand jury
- A management official or other employee of the contractor, subcontractor, r grantee who has the responsibility to investigate, discover, or address misconduct

If an employee believes that he or she has been subjected to a reprisal for the disclosure, he or she may file a reprisal complaint with the NASA OIGa

OIG determines whether an investigation is needed

If No

OIG determines the complaint is:

- Frivolous
- Fails to allege a violation of a prohibition under the statute
- Previously addressed by another federal or state judicial or administrative forum

IF No investigation is required OIG notifies complainant that no further action will be taken

If Yes

OIG initiates an investigation

OIG has 180 days to submit its report to the head of agency but can request an extension of up to 180 days with agreement from the complainant

OIG submits a report to:

- Head of agency
- Complainant
- Contractor, subcontractor, grantee or subgrantee

Head of agency determines:

- No reprisal
- Issues an order denying relief
- Reprisal

May issue an order granting relief

Head of agency has 30 days to make a determination and may request additional information

Data Table for Figure 2: NASA Office of Inspector General (OIG) Process for Investigating Reprisal Complaints

Reprisal complaint received

Complainant fills out whistleblower form to be identified as a whistleblower. This begins the 180 days investigation period. If complainant does not complete the form required, chooses not to continue the complaint, or the complaint otherwise does not qualify for protections, the investigation is closed

If complaint is NOT investigated OIG notifies complainant of decision not to investigate

If Complaint is investigated

- Over the course of 180 days, investigators gather evidence from the contractor or grantee which may include emails, human resource records, and interviews
- If the OIG determines the investigation will likely go longer than 180 days, the OIG asks for written agreement from complainant for an additional 180 days
- OIG transmits report of findings to agency head, contractor, and complainant

Data Table for Figure 3: Disposition of Reprisal Complaints Received by NASA's Office of Inspector General (OIG) from 2008 through June 2017

1) OIG reported whistleblower disclosures 277
2) Whistleblower reported experiencing reprisal, such as demotion Reprisal Complaints 48
    a) Ongoing investigations 5
    b) Investigated reprisal complaints 6
        i, Reprisal complaints substantiated by OIG 3
        ii. Reprisal complaints not substantiated by OIG 3
    c) Not Investigated, 37
        i. Determined disclosure not covered by statute 7
        ii. Determined the complaint was frivolous 27
        iii. Complaint addressed through other forum 3

Data Table for Figure 4: Timeline for NASA Administrator's Response for 5 Investigated Reprisal Cases Subject to Statutory 30-Day Final Determination from 2008 through June 2017

	Statute requirements	Total time elapsed for Administrator response to OIG findings	Agency response to OIG findings not documented
Reprisal case 5	30	80	
Reprisal case 4	30	80	
Reprisal case 3	30	35	
Reprisal case 2	30	239	
Reprisal case 1	30		243

AGENCY COMMENT LETTER

Text of Appendix II: Comments from the National Aeronautics and Space Administration

Page 1

February 13, 2018

Ms. Shelby S. Oakley Director

Contracting and National Security Acquisitions United States Government Accountability Office Washington, DC 20548

Dear Ms. Oakley:

The National Aeronautics and Space Administration (NASA) appreciates the opportunity to review and comment on the Government Accountability Office (GAO) draft report entitled, "NASA Contractor Whistleblowers: Steps Taken to Implement Program but Improvements to Timeliness and Guidance Needed" (GAO-18-262), dated January 18, 2018.

In the draft report, GAO makes three recommendations to the NASA Administrator intended to improve NASA's implementation of its contractor and grantee employee whistleblower program. Specifically, GAO recommends the following:

Recommendation 1: The Administrator should monitor, evaluate, and make appropriate corrective actions, such as a documented process, to ensure it reviews reprisal complaints in a timely manner.

Management's Response: Concur. NASA will develop a documented process to ensure it reviews reprisal complaints in a timely manner that ensure all parties due process rights are protected.

Estimated Completion Date: March 31, 2019

Recommendation 2: The Administrator should review NASA's guidance or develop other guidance, including defining major modification, to clarify when whistleblower protections are conveyed.

Management's Response: Concur. NASA will review existing procurement policy relative to when whistleblower protections are conveyed and provide clarifying guidance as appropriate.

Estimated Completion Date: October 31, 2018

Page 2

Recommendation 3: The Administrator should communicate whistleblower protections to grantees and sub-grantees and their employees.

Management's Response: Concur. NASA will update the NASA Grant and Cooperative Agreement Manual to communicate whistleblower protections to grantees and sub-grantees and their employees.

Estimated Completion Date: October 31, 2018

Once again, thank you for the opportunity to comment on the subject draft report. If you have any questions or require additional information, please contact Dave Barrett on (202) 358-2074.

Sincerely,
Robert M. Lightfoot, J.
Administrator (Acting)

In: NASA: Major Projects and Select Issues
Editor: Oliver Cummings
ISBN: 978-1-53614-817-6

Chapter 6

NASA POLICY ON THE RELEASE OF INFORMATION TO NEWS AND INFORMATION MEDIA*

SCOPE

This directive sets forth policy governing the release of public information, which is defined as information in any form provided to news and information media, especially information that has the potential to generate significant media, or public interest or inquiry. Examples include, but are not limited to, press releases, media advisories, news features, and web postings. Not included under this definition are scientific and technical reports, web postings designed for technical or scientific interchange, and technical information presented at professional meetings or in professional journals.

APPLICABILITY

(a) This policy applies to NASA Headquarters, NASA Centers, and Component Facilities.
(b) In the event of any conflict between this policy and any other NASA policy, directive, or regulation, this policy shall govern and supersede any previous issuance or directive.

* This is an edited, reformatted and augmented version of a NASA website article: https://www.nasa.gov/audience/formedia/features/communication_policy.html.

PRINCIPLES

(a) NASA, a scientific and technical agency, is committed to a culture of openness with the media and public that values the free exchange of ideas, data, and information as part of scientific and technical inquiry. Scientific and technical information from or about Agency programs and projects will be accurate and unfiltered.

(b) Consistent with NASA statutory responsibility, NASA will "provide for the widest practicable and appropriate dissemination of information concerning its activities and the results thereof." Release of public information concerning NASA activities and the results of NASA activities will be made promptly, factually, and completely.

(c) To ensure timely release of information, NASA will endeavor to ensure cooperation and coordination among the Agency's scientific, engineering, and public affairs communities.

(d) In keeping with the desire for a culture of openness, NASA employees may, consistent with this policy, speak to the press and the public about their work.

(e) This policy does not authorize or require disclosure of information that is exempt from disclosure under the Freedom of Information Act (5 U.S.C. § 552) or otherwise restricted by statute, regulation, Executive Order, or other Executive Branch policy or NASA policy (e.g., OMB Circulars, NASA Policy Directives). Examples of information not releasable under this policy include, without limitation, information that is, or is marked as, classified information, procurement sensitive information, information subject to the Privacy Act, other sensitive but unclassified information, and information subject to privilege, such as pre-decisional information or attorney-client communications.

RESPONSIBILITIES

(a) The Assistant Administrator for Public Affairs is responsible for developing and administering an integrated Agency-wide communications program, establishing Agency public affairs policies and priorities, and coordinating and reviewing the performance of all Agency public affairs activities. The Assistant Administrator will develop criteria to identify which news releases and other types of public information will be issued nationwide by NASA Headquarters. Decisions to release public information nationwide by NASA Headquarters will be made by the Assistant Administrator for Public Affairs or his/her designee.

(b) NASA's Mission Directorate Associate Administrators and Mission Support Office heads have ultimate responsibility for the technical, scientific, and programmatic accuracy of all information that is related to their respective programs and released by NASA.

(c) Under the direction of the Assistant Administrator for Public Affairs, public affairs officers assigned to Mission Directorates are responsible for the timely and efficient coordination of public information covering their respective programs. This coordination includes review by appropriate Mission Directorate officials. It also includes editing by public affairs staff to ensure that public information products are well written and appropriate for the intended audience. However, such editing shall not change scientific or technical data, or the meaning of programmatic content.

(d) Center Public Affairs Directors are responsible for implementing their portion of the Agency's communications program, adhering to Agency policies, procedures, and priorities, and coordinating their activities with Headquarters (and others where appropriate). They are responsible for the quality of public information prepared by Center public affairs officers. They also are responsible for the day-to-day production of public information covering their respective Center activities, which includes obtaining the necessary Center concurrences and coordinating, as necessary, with the appropriate Headquarters public affairs officers.

(e) Center Directors have ultimate responsibility for the accuracy of public information that does not require the concurrence of Headquarters. (See "Public information coordination and concurrence," section (d).)

(f) All NASA employees are required to coordinate, in a timely manner, with the appropriate public affairs officers prior to releasing information that has the potential to generate significant media, or public interest or inquiry.

(g) All NASA public affairs officers are required to notify the appropriate Headquarters public affairs officers in a timely manner about activities or events that have the potential to generate significant media or public interest or inquiry.

(h) All NASA public affairs employees are expected to adhere to the following code of conduct:

1 Be honest and accurate in all communications.
2 Honor publication embargoes.
3 Respond promptly to media requests and respect media deadlines.
4 Act promptly to correct mistakes or erroneous information, either internally or externally.
5 Promote the free flow of scientific and technical information.
6 Protect non-public information.

(i) All NASA employees are responsible for adhering to plans (including schedules) for activities established by public affairs offices and senior management for the coordinated release of public information.
(j) All NASA-funded missions will have a public affairs plan, approved by the Assistant Administrator for Public Affairs, which will be managed by Headquarters and/or a designated NASA Center.
(k) Public affairs activities for NASA-funded missions will not be managed by non-NASA institutions, unless authorized by the Assistant Administrator for Public Affairs.
(l) The requirements of this directive do not apply to the Office of Inspector General regarding its activities.

PUBLIC INFORMATION COORDINATION AND CONCURRENCE

(a) General. All NASA employees involved in preparing and issuing NASA public information are responsible for proper coordination among Headquarters, Center, and Mission Directorate offices to include review and clearance by appropriate officials prior to issuance. Such coordination will be accomplished through procedures developed and published by the NASA Assistant Administrator for Public Affairs.
(b) Coordination. To ensure timely release of public information, Headquarters and Center public affairs officers are required to coordinate to obtain review and clearance by appropriate officials, keep each other informed of changes, delays, or cancellation of releases, and provide advance notification of the actual release.
(c) All public information shall be coordinated through the appropriate Headquarters offices, including review by the appropriate Mission Directorate Associate Administrator and mission support office head, or their designees, to ensure scientific, technical, and programmatic accuracy, and review by the Assistant Administrator of Public Affairs or his/her designee to ensure that public information products are well written and appropriate for the intended audience.
(d) Centers may, however, without the full coordination of Headquarters, issue public information that is institutional in nature, of local interest, or has been deemed not to be a Headquarters release. (The Assistant Administrator for Public Affairs or his/her designee will determine which public information will be issued nationwide by NASA Headquarters.) These releases must be coordinated through the appropriate Center offices and approved by the Center Director and Center Public Affairs Director. The Center Public Affairs Director is required to provide proper notification to the NASA Office of Public Affairs, Headquarters, prior to release. (The Assistant Administrator for Public Affairs shall publish guidelines

for the release of public information that may be issued by Centers without clearance from Headquarters' offices.)

(e) Dispute Resolution. Any dispute arising from a decision to proceed or not proceed with the issuance of a news release or other type of public information will be addressed and resolved by the Assistant Administrator for Public Affairs with the appropriate Mission Directorate Associate Administrator, mission support office head, Center Director, and others, such as Center Public Affairs Directors, as necessary. However, the appropriate Mission Directorate Associate Administrator shall be the arbiter of disputes about the accuracy or characterization of programmatic, technical, or scientific information. Additional appeals may be made to the Chief of Strategic Communications and to the Office of the Administrator. When requested by a Center Public Affairs Director, an explanation of the resolution will be provided in writing to all interested Agency parties.

INTERVIEWS

(a) Only spokespersons designated by the Assistant Administrator for Public Affairs, or his/her designee, are authorized to speak for the Agency in an official capacity regarding NASA policy, programmatic, and budget issues.

(b) In response to media interview requests, NASA will offer articulate and knowledgeable spokespersons who can best serve the needs of the media and the American public. However, journalists may have access to the NASA officials they seek to interview, provided those NASA officials agree to be interviewed.

(c) NASA employees may speak to the media and the public about their work. When doing so, employees shall notify their immediate supervisor and coordinate with their public affairs office in advance of interviews whenever possible, or immediately thereafter, and are encouraged, to the maximum extent practicable, to have a public affairs officer present during interviews. If public affairs officers are present, their role will be to attest to the content of the interview, support the interviewee, and provide post-interview follow up with the media as necessary.

(d) NASA, as an Agency, does not take a position on any scientific conclusions. That is the role of the broad scientific community and the nature of the scientific process. NASA scientists may draw conclusions and may, consistent with this policy, communicate those conclusions to the media. However, NASA employees who present personal views outside their official area of expertise or responsibility must make clear that they are presenting their individual views – not the views of the Agency – and ask that they be sourced as such.

(e) Appropriated funds may only be used to support Agency missions and objectives consistent with legislative or presidential direction. Government funds shall not be used for media interviews or other communication activities that go beyond the scope of Agency responsibilities and/or an employee's official area of expertise or responsibility.

(f) Media interviews will be "on-the-record" and attributable to the person making the remarks, unless authorized to do otherwise by the Assistant Administrator for Public Affairs or Center Public Affairs Director, or their designees. Any NASA employee providing material to the press will identify himself/herself as the source.

(g) Audio recordings may be made by NASA with consent of the interviewee.

(h) NASA employees are not required to speak to the media.

(i) Public information volunteered by a NASA official will not be considered exclusive to any one media source and will be made available to other sources, if requested.

Preventing Release of Classified Information to the Media

(a) Release of classified information in any form (e.g., documents, through interviews, audio/visual, etc.) to the news media is prohibited. The disclosure of classified information to unauthorized individuals may be cause for prosecution and/or disciplinary action against the NASA employee involved. Ignorance of NASA policy and procedures regarding classified information does not release a NASA employee from responsibility for preventing any unauthorized release.

(b) Any attempt by news media representatives to obtain classified information will be reported through the Headquarters Office of Public Affairs or Installation Public Affairs Office to the Installation Security Office and Office of Security and Program Protection.

(c) For classified operations and/or programs managed under the auspices of a DD Form 254, "Contract Security Classification Specification," all inquiries concerning this activity will be responded to by the appropriate PAO official designated in Item 12 on the DD Form 254.

(d) For classified operations and/or information owned by other Government agencies (e.g., DOD, DOE, etc.), all inquiries will be referred to the appropriate Agency public affairs officer as established in written agreements.

PREVENTING UNAUTHORIZED RELEASE OF SENSITIVE BUT UNCLASSIFIED (SBU) INFORMATION/MATERIAL TO THE NEWS MEDIA

(a) All NASA SBU information requires accountability and approval for release. Release of SBU information to unauthorized personnel is prohibited. Unauthorized release of SBU information may result in prosecution and/or disciplinary action. Ignorance of NASA policy and procedures regarding SBU information does not release a NASA employee from responsibility for unauthorized release.

(b) Examples of SBU information include: proprietary information of others provided to NASA under nondisclosure or confidentiality agreement; source selection and bid and proposal information; information subject to export control under the International Traffic in Arms Regulations (ITAR) or the Export Administration Regulations (EAR); information subject to the Privacy Act of 1974; predecisional materials such as national space policy not yet publicly released; pending reorganization plans or sensitive travel itineraries; and information that could constitute an indicator of U.S. government intentions, capabilities, operations, or activities or otherwise threaten operations security.

(c) Upon request for access to information/material deemed SBU, coordination must be made with the information/material owner to determine if the information/material may be released. Other organizations that play a part in SBU information identification, accountability and release (e.g., General Counsel, External Relations, Procurement, etc.) must be consulted for assistance and/or concurrence prior to release.

(d) Requests for SBU information from other Government agencies must be referred to the respective Agency public affairs officer.

MULTIMEDIA MATERIALS

(a) NASA's multimedia material, from all sources, will be made available to the information media, the public, and to all Agency Centers and contractor installations utilizing contemporary delivery methods and emerging digital technology.

(b) Centers will provide the media, the public, and as necessary, NASA Headquarters with:

1 Selected prints and original or duplicate files of news-oriented imagery and other digital multimedia material generated within their respective areas.

2 Selected video material in the highest quality format practical, which, in the opinion of the installations, would be appropriate for use as news feed material or features in pre-produced programs and other presentations.
3 Audio and/or video files of significant news developments and other events of historic or public interest.
4 Interactive multimedia features that can be incorporated into the Agency's Internet portal for use by internal and external audiences, including the media and the general public.

News Releases Concerning International Activities

(a) Releases of information involving NASA activities, views, programs, or projects involving another country or an international organization require prior coordination and approval by the Headquarters offices of External Relations and Public Affairs.
(b) NASA Centers and Headquarters offices will report all visits proposed by representatives of foreign news media to the public affairs officer for the Office of External Relations for appropriate handling consistent with all NASA policies and procedures.

INDEX

L

M

N

O

P

R

S

Commercial Crew Program, Space Telescopes and the Global Positioning System

Editor: Graeme N. Lutz

Series: Space Science, Exploration and Policies

Book Description: NASA's Commercial Crew Program is a multi-phased effort that began in 2010. Across the phases, NASA has engaged several companies, using both agreements and contract vehicles to develop and demonstrate crew transportation capabilities.

Hardcover ISBN: 978-1-53614-980-7
Retail Price: $160